Sparrow

Animal
Series editor: Jonathan Burt

Already published

Ant
Charlotte Sleigh

Ape
John Sorenson

Bear
Robert E. Bieder

Bee
Claire Preston

Camel
Robert Irwin

Cat
Katharine M. Rogers

Cockroach
Marion Copeland

Cow
Hannah Velten

Crow
Boria Sax

Dog
Susan McHugh

Donkey
Jill Bough

Duck
Victoria de Rijke

Eel
Richard Schweid

Elephant
Daniel Wylie

Falcon
Helen Macdonald

Fly
Steven Connor

Fox
Martin Wallen

Giraffe
Mark Williams

Hare
Simon Carnell

Horse
Elaine Walker

Lion
Deirdre Jackson

Moose
Kevin Jackson

Otter
Daniel Allen

Oyster
Rebecca Stott

Parrot
Paul Carter

Peacock
Christine E. Jackson

Penguin
Stephen Martin

Pig
Brett Mizelle

Pigeon
Barbara Allen

Rat
Jonathan Burt

Rhinoceros
Kelly Enright

Salmon
Peter Coates

Shark
Dean Crawford

Snail
Peter Williams

Snake
Drake Stutesman

Spider
Katja and Sergiusz Michalski

Swan
Peter Young

Tiger
Susie Green

Tortoise
Peter Young

Trout
James Owen

Vulture
Thom Van Dooren

Whale
Joe Roman

Wolf
Garry Marvin

Sparrow

Kim Todd

REAKTION BOOKS

Published by
REAKTION BOOKS LTD
33 Great Sutton Street
London EC1V 0DX, UK
www.reaktionbooks.co.uk

First published 2012

Printed and bound in China by Eurasia

British Library Cataloguing in Publication Data
Todd, Kim, 1970–
Sparrow. – (Animal)
1. Sparrows. 2. Sparrows – Ecology.
3. Introduced bird – United States – History – 19th century.
4. Sparrows in literature.
I. Title II. Series
598.8'83-DC22

ISBN 978 1 86189 875 3

Contents

Introduction

The sparrow is a slight bird, small and dun-coloured, easily crushed. In the Bible it is the picture of worthlessness, only redeemed by the notice of God. 'Are not two sparrows sold for a farthing? and one of them shall not fall on the ground without your Father', Jesus tells his disciples (Matthew 10:29). In *Hamlet*, the prince mutters 'There is special providence in the fall of a sparrow.'[1] Holding his life cheap, Hamlet brushes aside premonitions of death and goes off to fight Laertes.

In a world fascinated by the predatory and breathtakingly beautiful, the sparrow is the type of the common and humble. There's something generic about it. Picture the basic bird, the stripped-down, super-efficiency model, and a sparrow probably comes to mind. The one that might be better alone in the hand than paired in the bush. The one whose song is mainly 'chirp'. The Hebrew word that gets translated to the English 'sparrow' means 'bird' in general, particularly a twittering one. The root of the Old English *spearwa* means 'flutterer'.[2] Its Latin name, *passer*, was adopted as the root of 'passerine', the name for the largest order of birds, all those that perch and sing.

But this sense of insignificance can be deceptive, both mythologically and biologically. In *Brut, or Chronicle of Britain* (*c*. 1200) by the poet-priest Layamon, sparrows are set alight, carrying fire and vengeance to a town, burning it to the ground. Japanese

House sparrow.

stories tell of sparrows, overlooked and underestimated, enacting vicious revenge. In a Grimm's fairy tale, a man who tramples a dog in the road is hounded by a sparrow to his death. The house sparrow was once a bird of Europe, the Middle East and parts of Asia, one that lived close to people in villages and by cottages. This gave rise to an image of the sparrow as dowdy and domestic. But as European culture spread, so did the sparrow, settling in Labrador, Tierra del Fuego, Hawaii and New Zealand, becoming a symbol of pestilence, urban ills and unwanted immigration.

These days, house sparrows are literally everywhere. They hop through snow-blown Churchill, Manitoba, on the coast of Hudson Bay, as polar bears range over nearby ice floes. They roost on mango trees in Brazil, following the Transamazon Highway snaking into the rainforest. They squabble in the Saudi Arabian desert. Their introduced range now extends to every continent but Antarctica. In some cases, people brought sparrows to new lands; in others they found their own way. The sparrow is the most widespread wild bird in the world. Its story seems to be that of the meek inheriting the earth.

Though fragile, it thrives in the least natural of environments. Outside a strip-mall cafe, a mother sparrow picks at croissant crumbs, followed by three unkempt adolescents shivering their wings and begging. She ignores them, probing deep between the tiles after a sesame seed. In an underground parking garage, high-pitched cheeps echo, amplified by the cement walls. They drag nesting material inside traffic lights. In the Detroit International Airport, a red train runs along an overhead track, a CNN commentator discusses legalizing marijuana on a large screen, passengers discard their Egg McMuffin wrappers and prepare to board. The only flecks of green in this landscape of nylon carpet and slick floors are potted trees decorating the waiting area. At

Sparrow on metal decoration.

the top of one, a house sparrow preens itself, rustling the leaves. The trees are plastic. The bird doesn't seem to care.

They are alternately loathed vermin and objects of delight. The collective noun used to describe them is 'host', as in angels or an army. The question of whether introduced house sparrows

Watanabe Seitei (1851–1918), *Sparrows Flying*, album leaf; ink and colour on silk.

should be encouraged or exterminated ignited the 'Sparrow War' in America in the 1900s. In parts of Asia, the tree sparrow plays the role of the house sparrow, living near people and eating their scraps. It stirs the same ambiguous emotions. In tenth-century Japan, court diarist Sei Shonagon wrote in her *Pillow Book* under the heading 'pretty things': 'A baby sparrow hopping towards one when one calls "chu, chu" to it; or being fed by its parents with worms or whatnot, when one has captured it and tied a thread to its foot.'[3] In 1950s China, Mao declared sparrows one of the 'Four Pests', along with flies, mosquitoes and rats, and ordered his people to annihilate them. He eventually relented, taking the sparrow off the list and replacing it with the bedbug.

Often even their biggest detractors find themselves grudgingly charmed. Elliott Coues, who led the charge against the bird in the Sparrow War and called the birds 'foreign vulgarians' and 'animated manure machines',[4] wrote: 'For myself, I "rather like" them too; they rather amuse me and interest me, and are not at all disagreeable, as long as I can keep their disastrous results out of mind.'[5] Samuel Christian Schmucker used the house sparrows introduced to the United States as a case study in natural selection in his book *The Meaning of Evolution*. He wrote of the house sparrow,

> All sensible bird-men must clearly acknowledge that he is a very undesirable citizen. I write the above sentence to show that I realize the whole duty of the bird-lover in the matter of the sparrow. This pestiferous creature should be exterminated by traps, by grain soaked in alcohol, or strychnia, by fair means or foul. But personally, I am taking no share in his destruction.[6]

Denis Summers-Smith, author of *The House Sparrow* and *The Sparrows* (which looks at all the house sparrow's close relatives), spent a lifetime studying the little bird and concluded it was almost appallingly successful. In his first book, he asks 'But do I love the house sparrow? That I find difficult to answer, though I do know that I should find life extremely dull without them as my constant neighbors.'[7]

Part of the attraction is their attitude. A house sparrow may be drab, hyperactive and tone deaf, but it doesn't feel bad about it. In some stories, the fact that sparrows hop rather than walk is a rebuke from God. But the bird is so chipper, the punishment must not be hard to bear. The combination of small and bold led Corbin Motors to name its electric car the 'Sparrow',

The Corbin 'Sparrow', a petite electric car, produced 1999–2002.

the U.S. Navy to call a surface-to-air missile the 'Sea Sparrow' and assassination squads in the Philippines to dub themselves 'sparrow units'.

The house sparrow (*Passer domesticus*), evolved near humans, nesting in their eaves, eating their fruits, seeds and crumbs. In the sunroom where I work, surrounded by windows, the sparrows couldn't be closer. A male chirps in the almost bare bush, still clinging to a few red leaves. Another lands on a nodding grass stalk and picks out the seeds. A third lights on the windowsill and cocks its head at me at my desk, peering in as I'm peering out. In the autumn, when they gather in sociable flocks, a hundred sparrows explode out of the garage when I open the back door.

Their habitat seems to be us, perhaps one reason we find them annoying and endearing. This sense of commonality has worked its way into the language: a person can be 'sparrow-mouthed', (wide-mouthed), 'sparrow-legged' (skinny) or 'sparrow-blasted' (astonished or cuckolded).[8] This closeness and availability made them a popular subject for scientific studies, and they have taught us about evolution, invasion and adultery.

But all this is just the house sparrow, one of many kinds of sparrow. The birds called 'sparrows' are split between two families. The first is the Old World sparrows in the *Passer* genus of the *Passeridae* family, which in addition to the house sparrow includes the Dead Sea sparrow, the Eurasian tree sparrow, the parrot-billed sparrow, the cinnamon sparrow and a handful of other species. These are deemed the 'true' sparrows, though North American birding websites assure visitors that it is the Old World bird which is an imposter, 'not really a sparrow'.[9] The second group is the New World sparrows in the *Emberizidae* family, which includes the white-crowned sparrow, the sooty fox sparrow and the saltmarsh sharp-tailed sparrow. Despite the name, the two kinds of sparrow are only distantly related. The Old World sparrows are tied to weavers and the New World sparrows are kin to buntings.

The sparrow families are incredibly diverse. Some New World sparrows, like the San Clemente sage sparrow, stake a claim to

Sea Sparrow RIM-7 missile launching from USS *Harry S. Truman*, May 2007.

Sylvester Dwight Judd, 'Lark Sparrow', from *The Relation of Sparrows to Agriculture* (1901): an American sparrow that didn't make up for the lack of the European house sparrow.

(*opposite*) Two American sparrows: Sylvester Dwight Judd, 'Field Sparrow' and 'Song Sparrow', from *The Relation of Sparrows to Agriculture* (1901).

remote islands while Old World sparrows, like the house sparrow, build their nests in Home Depot. Even so, our associations with the birds are often similar. As the dusky seaside sparrow, a New World species, headed for extinction, newspaper headlines asked: 'Whose eye is on the sparrow?',[10] referencing the sparrow mentioned by Jesus, most likely an Old World house sparrow. The American poet Emily Dickinson wrote: 'God keep His Oath to Sparrows / Who of little Love – know how to starve –'[11] and

> Her heart is fit for home
> I – a Sparrow – build there
> Sweet of twigs and twine
> My perennial nest.[12]

J.L.R.
J.L.R.

Dickinson writes about the literary house sparrow of the Bible and the Romantic poets, not any of the New World species – like the field sparrow or vesper sparrow – that she might have glimpsed on a branch in Amherst. It's the one God promised to pay attention to, even if not much, the one that nests so close. In both cases, the poet identifies with the bird. Other animals, known for fierceness or fidelity, are often what we aspire to. In poem after poem by Dickinson and others the sparrow is, despite ourselves, what we are.

The subject of this book is this idea of the sparrow, that flitting creature of imagination – trickster, lecher, rebel, innocent, pest – as much as the biological entity that took advantage of the nooks in our buildings and our spilled grain. It will look at the bird's evolution, its communities and the house sparrow's recent, mysterious vanishing from European cities, but will also raise the question: what do we make of almost nothing? How do we treat the insignificant, the overlooked, outside our windows and within ourselves?

1 Little Brown Jobs

It's hard to generalize about house sparrows. One of the reasons they're so globally successful is they never met a statement about their biology or physiology they didn't challenge. They are primed to adapt.

House sparrows appear to have the ability to live everywhere humans have set foot (except the moon and Antarctica – so far). In 1915, a newspaper reported that a house sparrow was living 229 metres (750 feet) underground in a Scottish coal mine. For company, the sparrow had a mouse, a brown rat, a slug, some beetles, flies and a pit flea.[1] Reports continued in the 1950s and '60s, claiming sparrows in mines in Northumberland and Durham. Natural historian Denis Summers-Smith read these with interest, but it wasn't until the mid-1970s that he was able to visit the Frickley coal mine in Yorkshire and see for himself.

Three house sparrows – two males and a female – had abandoned the sparrow community in the mine buildings at the surface in order to live 640 metres (2,100 feet) down. The miners fed them, and they persisted, staying near the electric lights and out of the larger maze of stone. Summers-Smith speculated that they came down the 'skip shaft', a hole used to haul carts of broken rock in and out of the mine. They built a nest on a roof support and hatched out three young, but the young disappeared.

House sparrow (*Passer domesticus*).

On Summers-Smith's trip down, he saw just the male and deemed him 'in excellent condition'.[2]

This subterranean life is particularly surprising for a creature so swayed by light. House sparrows, like many other birds, are governed by seasonal shifts. Their body and behaviour change depending on the length of light in a day. For example, at the start of spring, the increase in sunshine triggers the creation of sperm and the development of eggs. The testicles get 100 times bigger; the ovaries about 50. For the rest of the year, these parts are neatly tucked away.

As the days get longer, males scout for a promising place to build a nest. Though house sparrows got their common name nesting on roofs, they will build anywhere with a suitable crevice: drain spouts, traffic lights, furled sails, the raised plastic curves of *S* and *O* shapes on shop fronts. Human structures are not the only ones house sparrows will adopt. They've built alongside the nests of osprey, red-tailed hawk and the common pariah kite.[3] Apparently a host with a sharp beak and talons doesn't faze them. One ornithologist found house sparrows raising chicks in the side of a Swainson's hawk nest in an elm, while a hawk nestling tore apart baby cottontail rabbits a few feet away.[4]

The nests themselves are messy heaps of grass or straw, a few twigs. As with so much about sparrows, nest building codes are flexible. The domed structure often has an entrance on the side, leading to a enclosure lined with hair, wool or feathers (sometimes yanked from other birds while still attached). The sparrows can incorporate greenery, like wild carrot, maybe to help ward off parasites. Sparrows are not always that popular with competing species, but, when not fighting over a mate, they enjoy each other's company. Twenty or so nests can be built very close to one another, making a sparrow apartment complex.

With a nest site secured, a male without a mate starts calling 'chirrup'. If a male shows up, the chirper defends his territory vigorously, even if it is only a few feet on either side of a given

Sparrow on a McDonalds's sign, Ohio, June 2008.

lamppost. He flies in, claws extended, beak low, wings canted back, pecking and hopping and fluttering like a miniature marionette. The viciousness of the fights has commonly been remarked upon, as well as the bird's seeming hair-trigger: 'Jealousy is certainly the *vera causa* of the sparrow's irritability and pugnacity. This feeling is so deeply ingrained into its very being that the slightest cause will evoke it', wrote Thomas Gentry, a nineteenth-century observer, in *The House Sparrow at Home and Abroad*.[5]

If a female responds by crouching with her tail up, calling 'quee' and drooping her wings, the male drops his wings as well and 'shivers' them, inviting her over. When they mate, often the male hops on the female's back ten times or more. They often do this on the porch, on the sidewalk, in the playground, in the road, in full view, giving them a lecherous reputation.

Commentators as early as Aristotle included cock sparrows among the 'salacious animals and such as abound in seed'.[6]

Sparrows squabbling in the street.

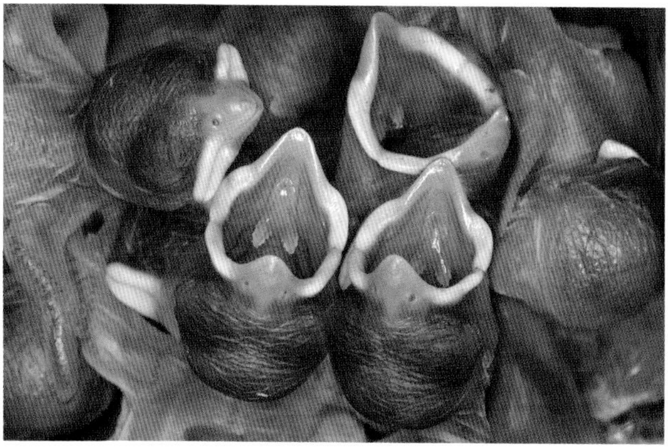

Newly hatched sparrows.

In his essay 'On Longevity and Shortness of Life', where he proposes that warmth and moisture are necessary to animal existence, he gives all this wanton spending of warm, moist seed as the reason female sparrows live longer than males. Pliny the Elder, the Roman natural historian, echoes Aristotle in his *Natural History*, published in AD 77–9, reinforcing the ties between lust and death. He calls the randy birds 'short-lived in the extreme', adding 'It is said that the male does not live beyond a year.'[7]

Because sparrows were so lusty, those suffering from impotence were advised to eat them. And they did: sparrow brains, sparrow rumps and the whole roasted bird were taken as aphrodisiacs. In books of occult philosophy written by Henry Cornelius Agrippa, the birds' obsession with mating takes a gentler turn. According to Agrippa, witches use the blood of sparrows in their love potions. Rings left in the nests of sparrows bring the wearer love.

Historically, many observers thought of Old World sparrows as monogamous with strong pair bonds lasting from season to

Oliver Herford (1863–1935), *Sparrow Feeding Young with Spoon*, drawing.

season. A week or so after mating, the female lays between two and five whitish eggs with black speckles. A pair can raise up to four broods over the course of the summer. Both male and female sit on the eggs. Both bring the nestlings grasshoppers, spiders and caterpillars and shove them in the waiting mouths. They seem model parents.

But when DNA testing became available in the late 1980s, making it possible to learn the precise parentage of chicks in a nest, all sorts of alternate configurations were revealed. In one study of about 60 house sparrow nests, 8 per cent of chicks had a parent that didn't belong to the nest. One male, a serial monogamist, nested with three different females in succession. Some are bigamists, mating with two females at once. One female built a nest with her son. If incest wasn't bad enough, she cheated on him: one of the nestlings belonged to a neighbour.[8]

These complex domestic arrangements can lead to behaviour that would have shocked those touting sparrow's parenting skills. A recent article by José Veiga in *Animal Behaviour* reads

A careful female diligently feeds a nestling.

more like a crime novel than a scientific study. Veiga set up nest boxes in the Guadarrama mountains of Spain and noticed that the second largest cause of death of nestlings (after being snatched by predators) was infanticide. Both sexes were guilty, with ample motive and opportunity, and Veiga either observed the killings first hand, or pieced together what had happened when he found beak-shaped puncture marks on the bodies. Sometimes a male who'd lost his mate would go into a nestbox, peck the chicks to death, throw them out and mate with the female. In other instances a male had two nests, one with a primary partner, whose nestlings he'd feed, and another with a secondary female whom he ignored. The secondary then killed the young of the primary, earning her the help of the male at her nest. Veiga concludes that all this violence reinforces monogamy as a desirable strategy.[9]

In the autumn, flocks of sparrows gather at harvest fields to chase stray grains of wheat or corn. As the days shorten, the testes and ovaries shrink, freeing the birds of extra weight. Most don't migrate, taking advantage of stored food and sheltered roosting spots to help them through the winter. Through the cold months, they hang out in loose groups, squabbling over seeds, taking dust baths, chirping in a rough chorus, taking refuge in roost sites when it gets too chilly and waiting out the snow until it's time to think about nesting again. Summers-Smith notes their easy life, even when living at climate extremes.

Sparrows gather around a fancy birdhouse.

Flock of sparrows in a tree top.

He wrote in *The House Sparrow* that, outside the breeding season, 'House sparrows appear to have plenty of spare time.'[10]

When it comes to eating, like nesting, house sparrows are innovative risk-takers. Though they give their young insects during the first few weeks of life, mostly they eat seeds or, if they live in the right neighbourhood, the best scraps they can find. They are not picky. Tests of the stomach contents of birds in Pennsylvania found corn, wheat, oats, millet, sunflower seeds and elm seeds; weeds like wood sorrel, panicgrass, bristlegrass and starwort; plant parts from leaves to anthers; Japanese beetles, gypsy moth caterpillars, insect eggs, spiders, aphids, mites, ticks, flies, tapeworms, ants, lice, fleas and bread. And that's just the healthy stuff.[11]

In New Zealand house sparrows learned to trigger the automatic doors at a bus station, giving them access to the buttery remains at the cafe inside. They perched on the sensor and dipped their heads in front of it, or flew right through the green beam of light.[12] When cars replaced horses, observers predicted a steep decline in the sparrows that spent so much time picking oats out of manure. Some vanished, but others learned to scavenge crushed insects from automobile grilles. As humans develop novel foods such as chilli cheese fries and Cool Ranch Doritos, house sparrows are ready to devour the crumbs.

They are also accomplished thieves, exhibiting a doggedness that can look, to the anthropomorphizing eye, like bravery or corruption. An 1885 newspaper article in Troy, New York, quoted biologist C. Hart Merriam condemning sparrows for taking grain: 'The most desperate criminal we have had to deal with is the pugnacious little English sparrow . . . He has been caught in the act of committing most terrible depredations, and upon

A sparrow stealing spaghetti off a plate.

Neeta Madahar, *Sustenance 51*, 2003, photograph.

examination his stomach has been found to be overloaded with stolen provisions.'[13]

They take flies from spider webs and worms from robins. In Minnesota, gangs of house sparrows have been reported to steal katydids from great golden digger wasps. The wasps nest in the dirt, excavating a tunnel and laying an egg in the burrow at the end. The female paralyses and captures an insect such as a katydid, often larger than she is, then lugs it back to the nest, sometimes flying and sometimes dragging her prey to the burrow, where it will feed the larva when it hatches. Sparrows will fly at the wasp, causing her to drop the insect, or harass her on the ground until she lets it go, or grab it when she leaves it at the mouth of the tunnel and goes down to check on the egg. Wasps lose up to a quarter of their prey this way. And sometimes, after a sparrow claims the katydid, the other sparrows steal it.[14]

The Old World sparrows, the genus containing house sparrows, are characterized by their nests, those jumbles of sticks with a domed roof and a lining of feathers or soft grass. Often the male and female have different colouring. In more than half the roughly 27 *Passer* species (ornithologists don't agree on the number),[15] the male has a black patch spreading from chin to chest, called a 'bib' or 'badge'. His bill turns black in the breeding season. (One of the reasons Pliny thought females lived longer than males was that in some seasons he didn't see any birds with black bills at all.)

In shape, they are variations on a theme, as if a sculptor had scooped up a palm-sized chunk of clay and moulded a simple bird, notching the tail, smoothing a plump chest, using his fingers to squeeze the beak to a chunky point. And then, when he was done, he liked it so much he made another, and another. With the flock complete, he took a limited palette of white, cream, black, grey, buff, chestnut and a startlingly bright gold, and let his imagination roam over the males. He made the whole chestnut sparrow the warm brown of hot cocoa, painted the golden sparrow the colour of lemonade and gave the White Nile rufous an angry yellow eye. He ran a thick brush of grey down the back of the Somali sparrow, and daubed a thin short beard like a sheik's on the male desert sparrow. Females, on the other hand, don't even have these modest distinguishing features, being mostly covered in brown flecks.

The genus extends from Europe through Africa to Asia. Though they look similar and tolerate people (all but three species can nest in buildings), the behaviour of the Old World sparrows differs. The golden sparrow roosts in flocks of hundreds in Senegal; the Sind Jungle sparrow of India prefers groups of five or six. The Dead Sea sparrow builds nests in trees (unlike most of the others, which like holes) in an area so hot, it scarcely

A male cape sparrow (*Passer melanurus*), Namibia.

needs to sit on the eggs at all. Willow sparrows are nomadic, chasing their food, while most other species stay in one place.

The Eurasian tree sparrow (*Passer montanus*) is the house sparrow of the east. Though it occurs in patches in Europe, it has Japan and southeast Asia all to itself. An Everest expedition found one at 3,960 metres (13,000 feet) in the Himalayas.[16] In this species, both the male and female have black bibs (this must lead to some confusion when house and tree sparrows interbreed, which they do, but apparently they can work it out). A black dot is daubed in the middle of the white patch on its cheek, giving it the appearance of a clown. When house sparrows are absent, tree sparrows live closely with people; where their ranges overlap, the house sparrow dominates, shoving the tree sparrows to more rural areas. For example, the tree sparrow

American tree sparrow (*Spizella arborea*) with chicks.

was introduced to St Louis in 1870, but its spread was checked when the house sparrow arrived not long after. Today the tree sparrow can still be found in St Louis, while the house sparrow has moved in from New York to California.[17] To confuse things further, a New World species is called the 'American tree sparrow', and bears no relationship to the Eurasian bird.

The sparrows of the African tropics are the least like the rest of the genus: the sexes look alike, they nest in grasslands, they aren't as fond of people. These Old World sparrows, called 'grey-headed sparrows', may have been the source of all the others. According to Summers-Smith, during the Pleistocene this not-very-fancy-looking bird was eating grass seeds and building domed nests in trees. One race, in which the male developed a black bib, followed the Nile and Rift Valleys to the Mediterranean about a million years ago. Glaciers, building and receding, divided that single type into nine separate species.[18]

'House Sparrow', from *Birds of Europe* by John Gould (1832–5).

The house sparrow got a foothold in the Fertile Crescent as people developed agriculture (all the better to steal grain) and erected buildings (all the better to nest in). The shelter of an

enclosed house and a stored food source enabled the house sparrow to stick around all winter and give up its migratory ways. Archaeologists have found fossils of house sparrow ancestors in a cave near Bethlehem and in a cave on Mount Carmel, a mountain singled out in the 'Song of Songs' for its beauty. Given the dates of these sites (400,000 years ago and 65,000 years ago) these sparrows may have been pestering humans when humans were barely human.[19]

More recent archaeological sites have offered further evidence of the relationship. In Sweden, archaeologists digging up an area that was inhabited 3,000 years ago uncovered wells, pottery fragments and the remnants of long houses. It was a rich find, and they didn't want to miss anything. Using a sieve, they sifted tiny bits of wing and leg from the soot and charcoal in cooking areas and refuse pits. There, among the heaps of Bronze Age trash, were the bones of house sparrows.

Some lay scattered by the fire. One was embedded in the clay floor. Sparrows must have been perching by the cooking pots, or being cooked in them. This was before Northern Europeans kept chickens, but at roughly the same time they started using domesticated horses. The lives of humans and sparrows in Sweden were woven together, even as people were just figuring out how to fuse copper and tin.[20] House sparrow remains also appeared in central Spain during the Iron Age, before the donkey, around the same time as the house mouse.[21]

This hanging out with humans may have created the sparrow as we know it, making *Passer domesticus* into the bird we recognize today. The birds are 'commensals', a relationship where one species benefits and the other species isn't harmed, like a barnacle hitching a ride on a whale. House sparrows don't just have a high tolerance for living near humans; they prefer it. Given the option between nesting on a clip holding up a gutter

Fox sparrow (*Passerella iliaca*).

spout or in a hole in a tree, they choose the gutter. In hot climates, like India, they nest not just on the houses, but inside the cool rooms, tucking messy heaps of grass on beams and the tops of blinds. They also like our toys. In one experiment, house sparrows approached food more quickly when it was placed next to a rubber ball and a plastic lizard than when the food was by itself.[22] They are hardwired for the Happy Meal.

Across the Atlantic, another group of birds was forming. There are about 73 genera in the *Emberizidae* family (here too, ornithologists keep changing their minds on the exact number and who belongs where), ranging from the grassquits to the towhees, juncos and buntings, fox sparrows and crowned sparrows.[23] The family is made up of 'American sparrows' and buntings. There are so many, of such diverse kinds, yet they are so hard to tell apart that they are referred to by some bird watchers as 'little brown jobs', or 'LBJS' for short.

American sparrows eat seeds and mostly create cup-shaped nests. They like shrubs and grassland and get much of their food from, and often build nests on, the ground. They range from the

Golden-crowned sparrow (*Zonotrichia leucophrys*).

secretive LeConte sparrow of the central plains that likes wet meadows and peat to the rufous-collared sparrow, which doesn't mind the city and is spread from Central America to the tip of Chile. The finches Darwin found in the Galápagos, with beaks tailored to seeds of every shape, are members of the *Emberizidae* family. They often have streaks on their chests and stripes or other decoration on their heads.

Most are found in North and South America, though one genus, the *Emberiza*, lives in Europe and Asia, ranging from Scandinavia to Turkey, Mongolia and Japan. (Commonly called the 'Old World Buntings', they could also be called the 'Old World New World Sparrows'.) This broad sweep of species, genera and families makes one wonder how such different kinds of birds all ended up with the same name.

The first reports by European settlers in the New World had little time for mundane critters. They were too busy with the fantastical: birds that nested in chimneys and threw down one of their young in gratitude, birds whose beaks could cure toothache, birds that spent all winter 'starke naked' on the nest.[24] Even when it came time to make a more cold-eyed assessment, it was hard to spare a glance for small, brown birds, when faced with white-headed eagles able to carry off a pig, tasty wild turkeys and shining hummingbirds that could disappear entirely into a blossom, where children could catch them.

Even so, it was clear to the colonists early on that there weren't any sparrows like the ones back home. John Josselyn, who travelled through Maine and Massachusetts in the mid-1660s, lists in his *New Englands Rarities Discovered* all the species the New World lacks. 'Now by what the Country hath not, you may ghess at what it hath; it hath no Nightingals, nor Larks, nor Bulfinches, nor Sparrows, nor Blackbirds, nor Magpies, nor Jackdawes, nor Popinjays, nor Rooks, nor Pheasants, nor Woodcocks, nor Quails, nor Robins, nor Cuckoes, &c.'[25] Many birds, at least in their familiar forms, were missing.

John Lawson, a young Englishman surveying the interior of North Carolina in 1701, reveals most clearly how the New World, with no *Passer* birds, became populated with sparrows. He itemizes birds in his book *A New Voyage to Carolina*, commenting, 'Sparrows here differ in Feather from the *English*. We have several Species of Birds call'd Sparrows, one of them much resembling the Bird call'd a *Corinthian* Sparrow.'[26] Then he adds later, of the 'reed sparrow': 'This nearest resembles a Sparrow, and is the most common Small-Bird we have, therefore we call them so. They are brown, and red, cinnamon Colour, striped.'[27] It's not just the smallness and brownness that makes a sparrow, but commonness, one of the bird's most distinguishing

Thomas Bewick (1753–1828), *Java Sparrow*, pen and grey ink and watercolour. The Java sparrow got its common name because of its rough size and shape.

traits. European settlers to North America, Australia and New Zealand found red-breasted birds wherever they went and called them 'robins.' (The American robin is a thrush and is easily twice the size of England's robin redbreast.) In the same way, they named chipper brown birds 'sparrows' in a nostalgic approximation.

While Josselyn charted the paucity of American fauna, Lawson emphasized the opposite, declaring at the top of his list 'Birds in America more beautiful than in Europe'.[28] They were engaging in a debate that would continue to frame the way sparrows were viewed for centuries after. Was America deficient, missing all the much-loved creatures that played starring roles

Anon., *Java Sparrow*, Guangzhou, 1800–1830, watercolour and ink on paper.

in stories and fables, enriching the lives of Europeans? Or were these new species opening a whole new realm of possibility, of launching pads for the imagination?

Later accounts by more careful observers began to weave the term 'sparrow' into the fabric of Americans' understanding of their wildlife. In his *Natural History of Carolina, Florida and the Bahama Islands* (1731), Mark Catesby painted a 'Bahama sparrow' and a 'little sparrow' (still unidentified).[29] William Bartram, an explorer of the southeast, writes of white-throated sparrows and field sparrows. By the time Audubon put together his outsize *Birds of America* in 1827, American sparrows were well established, though he termed many birds that would later be

Mark Catesby, 'The Little Sparrow', from *The Natural History of Carolina*, vol. I (1731), etching.

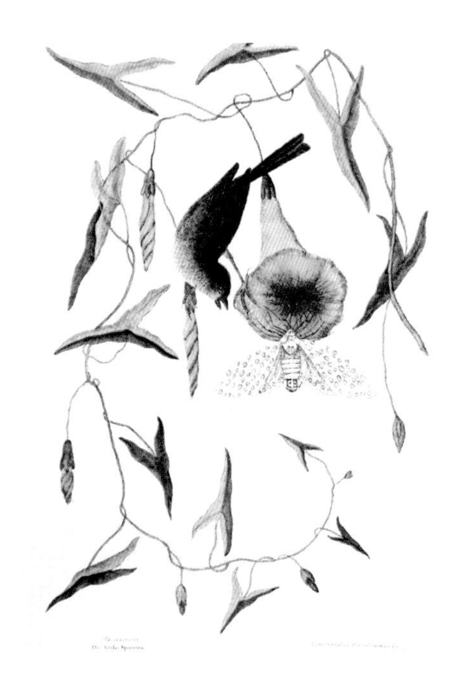

called sparrows 'finches'. He included illustrations of swamp sparrows and tree sparrows, among others.[30]

Unlike the true sparrows, which go 'chirrup', 'quee', or occasionally 'chree' when they are frightened, many of these New World birds are known for their complex, evolving music. The male song sparrow (*Melospiza melodia*) can sing a dozen different songs, and females prefer both songs with more learned parts than hardwired parts and males with larger repertoires over those with only a few tunes at their disposal. This bird's music strikes the human ear as particularly ecstatic and blissful, and it was one of the first of the New World species to be singled out by poets and essayists. Celia Thaxter in *The Atlantic Monthly* in 1873 described

John James Audubon, 'Swamp sparrow' (*Melospiza georgiana*), from *The Birds of America* (1827–30).

John James Audubon, 'Song sparrow' (*Melospiza melodia*), from *The Birds of America* (1827–30).

Song sparrow (*Melospiza melodia*).

His strain of rapture not to be suppressed . . .
That song of perfect trust, of perfect cheer,
Courageous, constant, free of doubt or fear.[31]

Because there are so many variations on the theme, both in the Old World and the New, it must be a good thing for a bird to be small, streaked with beige, grey and white, and live on seeds. The pattern evolved repeatedly, spanning continents and families. Seeds are easy to come by. Nondescript birds are hard to see, difficult to distinguish from each other and the landscape. They are the colours of sunlight through dry branches, the points of bramble thorns and fingers of shade, and can disappear with ease. As a model for survival, the Little Brown Job is advantageous and enduring.

2 Sold for Two Farthings

By far the most familiar literary sparrow lives in the New Testament. In the Gospel of Matthew (10:29–31), Matthew details Jesus choosing his twelve disciples and giving them their mission. Jesus advises them how to proceed, warns them they will be hated, tells them to run when chased. It is a worrisome task, but Jesus says not to be afraid, that they are watched over and cared for. He adds:

> Are not two sparrows sold for a farthing? and one of them shall not fall on the ground without your Father.
>
> But the very hairs of your head are all numbered.
>
> Fear ye not therefore, ye are of more value than many sparrows.

Luke tells a variation of the same story, though the sparrows have got even cheaper: 'Are not five sparrows sold for two farthings, and not one of them is forgotten before God?' (12:6–7)

Sparrows are as plentiful as hair, almost uncountable by humans, yet accounted for by God. Jesus' advice establishes both that even the smallest things are worthy of attention, and that not all life is equal. Matthew reinforces the message of God's care, and of human measure, when he records Jesus

Franz Marc, *Dead Sparrow*, 1905, oil on wood.

telling his followers not to worry about their next meal: 'Behold the fowls of the air: for they sow not, neither do they reap, nor gather into barns; yet your heavenly Father feedeth them. Are ye not much better than they?' (6:26). The passage doesn't mention sparrows, but they are often read into it, as in the painting *The Gleaners* by Stephanie Frostad where field sparrows eat dropped grain. In her *Hospital Sketches,* nineteenth-century American author Louisa May Alcott ties the fallen sparrow directly with the bird fed by God: 'I stopped away altogether, trusting that if this sparrow was of any worth; the Lord would not let it fall to the ground. Like a flock of friendly ravens, my sister nurses fed me, not only with food for the body, but kind words for the mind.'[1] This Christian bird, a meditation on a low-status life, flits its way through poems, plays and songs for the next 2,000 years.

The mention in Matthew raises the question: who would buy a sparrow, and why? It's less surprising that they are such a

J. H. Gurney, frontispiece from *The House Sparrow* (1885).

bargain, than that they have any monetary value at all. Shonagon in *The Pillow Book* writes about keeping a sparrow on a string and feeding it, so in some cultures it had value as a pet. One of the few European paintings featuring a sparrow, *Madonna del Passero* by Guercino, shows just this relationship. In this Italian painting, completed around 1615, Mary holds a naked infant Jesus on her lap and a house sparrow on a string on her finger. The faces of both figures are in shadow and they stare intently at the little bird who, in full light, stares back. The simplicity

Stephanie Frostad, *The Gleaners*, 2008, oil on paper.

Joseph Stella, *Lilies and Sparrow*, c. 1920, drawing, coloured pencil, silverpoint and white gouache.

and stillness seem as though they will be disrupted at any moment as the baby reaches for the bird or the sparrow tries to fly away. It is at once a mother and her son finding delight in nature, as Shonagon does, and holy figures with their eyes on the insignificant.

In later years, farmers sick of the flocks raiding their grain would pay bounties for dead birds. But if there was any real market for sparrows in Jesus' time, it was probably as food. A tenth-century cookbook from Baghdad records a recipe for a sparrow omelette:

Guercino, *Madonna of the Sparrow*, 1618–20, oil on canvas.

> Clean the sparrows and fry them in oil and a little salt until they brown. Take 10 eggs, and beat them with a little

black pepper and chopped cilantro. Pour the egg mixture on the sparrows, fry them until they are done, and serve them, God willing.[2]

In the eighteenth and nineteenth centuries, English and American cookbooks offered recipes for sparrow dishes as well. The several dozen required for a pie would take a lot of plucking, but when laid into a buttery crust, sprinkled with sage or dotted with mushrooms, it might be worth the effort. Modern-day hunter-gatherers also find the little bird tasty. A friend with a sparrow trap in his Brooklyn backyard strings them together to make shish kabob. Each alone would be a very small mouthful, probably why they were so cheap.

Sparrows also appear in Old Testament psalms, associated with human architecture. In Psalm 84, the sparrow and swallow find a place to nest in the house of the Lord, as they did so often in the houses of his followers. In Psalm 102, a song of suffering, the speaker feels abandoned and far from God (6–7). In some of the Bible's most stirring bird imagery, he says:

Jesus, watched by Joseph and three companions, commands the sparrows he made on the Sabbath to fly away, from *Gesta infantiae salvatoris* (1470–80).

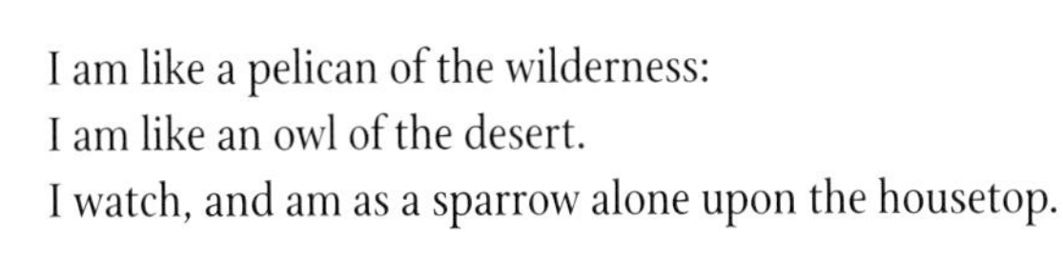

I am like a pelican of the wilderness:
I am like an owl of the desert.
I watch, and am as a sparrow alone upon the housetop.

This is an unusual pose for a sparrow. The bird is almost never alone, preferring to travel in a raucous cloud of friends, making the image in the psalm all the more stark, the loneliness more acute.

In one apocryphal tale, the disciples themselves are sparrows, formed by Jesus out of clay. In the Gospel of Thomas, a collection of stories about Jesus as a boy, the five-year-old Jesus makes twelve mud sparrows from a pool created by a rainstorm.

Passer, the Sparrow, in a bestiary from c. 1325–50.

Other children tell on him for breaking Sabbath, and Joseph scolds him. Jesus tells the sparrows no one will harm them. He claps his hands, and they fly off. In another story, a young Jesus laughs as twelve sparrows squabble over corn he has scattered, disturbing a lecturing school teacher. The sparrows and Jesus' interaction with them thwart the established order. They are playing when they should not be, learning a different kind of lesson.[3]

In these early writings, it is unclear whether the bird referred to is actually a sparrow, or just a tiny bird. The original language is imprecise and the sparrow character in the Bible has few of the biologically built character traits of the actual house sparrow. It isn't raucous or resourceful or in love with a mob. In these biblical strands, the sparrows are are like us, just a bit less, whether lonesome, happy in God's house, worthy of divine attention. They are stand-ins for humans. God makes man out of clay. His son, a child testing his abilities, makes sparrows. They are the practice miracle.

But in a story from the Book of Tobit (2:7–10), the bird under consideration is surely a house sparrow. One hot night, Tobit, a generous and righteous man, buries a fellow Jew, going against the king's orders. Then he falls asleep on a bench by a wall, neglecting to cover his face. As he sleeps, droppings from sparrows on the wall fall into his eyes, forming a white film and blinding him. Who else but house sparrows would be loitering in town, annoying the tired and pious with their excrement?

This notion of God's interest in the sparrow is picked up and repeated in countless sermons and poems, but Shakespeare gives

After Maarten van Heemskerck, *The Story of Tobias: Tobit Blinded by Sparrow Droppings*, 1550, woodcut.

it new life and takes in a new direction. The lines from Matthew are amplified and shaded darker by Hamlet. Challenged to fight by Laertes near the end of the play, the prince tells his friend Horatio that he is apprehensive, even though he thinks he is the more skilled swordsman. Horatio urges him to delay, and Hamlet replies: 'Not a whit, we defy augury: there is special providence in the fall of a sparrow. If it be now, 'tis not to come; if it be not to come, it will be now; if it be not now, yet it will come – the readiness is all.'[4]

In Hamlet's view, God's attention to detail is a prison. If Hamlet, like a sparrow, is meant to fall, he will fall. He can't escape from God's plan, so his decision to duel or not doesn't matter. The passage is often read as a statement of hope, that even this small death has meaning. But Hamlet doesn't see his fall as significant, as much as unavoidable. Like Emily Dickinson in her poetry, he is the sparrow, this smallest of sparks of life. God keeps track of him, but isn't invested in his welfare.

Hamlet mentions the sparrow in terms of augury and, like many other birds, it had been used to divine the future. According to Agrippa, the sparrow, with all her fecundity, is a good omen to those who want children, but is bad news to someone running away as 'she flies from the hawk and makes hast to the Owl, where it is in as great danger'.[5] In the *Iliad*, Odysseus uses a sparrow prophecy to stem a rebellion among the Greeks, who are homesick and tired of the war. He reminds them that when their ships were gathered at Aulis, a snake slithered from the altar during a ritual sacrifice to the gods. It crawled into a tree and ate a nest of eight sparrows and their panicked mother before being turned to stone. Odysseus explains this as a statement of the gods that after nine years fighting Troy, the Greeks will finally take the city.[6] Here prophecy is a promise, not the divine micro-managing of Hamlet's interpretation.

Another sparrow character also appears in Shakespeare's work. This is the dunnock or hedge sparrow (*Prunella modularis*), which, like the New World sparrow, is not really a sparrow. Cuckoos parasitize dunnock nests, sneaking in and laying several of their own eggs, then fly away, leaving the hedge sparrow parents to raise them, often at the expense of their own chicks. This small detail, noted by Aristotle, who wonders if the parasitized birds ignore their own young because they are dazzled by the cuckoo's beauty, defines the literary personality of the species.[7] Hedge sparrows are portrayed as pathetic dupes (a nickname was 'foolish sparrow'), rarely ever the case for the house sparrow.[8]

In *King Lear*, Lear's daughter Goneril complains to the King, who has come to stay with her, that his Fool is unruly. The Fool comments that 'the hedge-sparrow fed the cuckoo so long, that it had it head bit off by it young, so out went the Candle, and we were left darkling'.[9] The daughter is so ungrateful that it is

'Hedge Sparrow', from *British Birds* by F. B. Kirkman and F.C.R. Jourdain (1950).

Virgil Solis, 'The Cuckoo in the Sparrow's Nest', from Nicolaus Reusner's *Emblemata* (1581).

as if she is from another species, one that had no compunction about consuming her parent.

This theme of ingratitude appears again in *Henry IV, Part 1*. Worcester complains to the King that he forgot his promises to those who nurtured and protected him in his youth:

> And being fed by us you us'd us so
> As that ungentle gull, the cuckoo's bird,
> Useth the sparrow; did oppress our nest;
> Grew by our feeding to so great a bulk
> That even our love durst not come near your sight
> For fear of swallowing; but with nimble wing
> We were enforc'd, for safety sake, to fly
> Out of your sight . . .[10]

Once gone, they raise an army to defy the King, and the much-abused hedge sparrow turns warrior. After the revolution

Hedge sparrow from W. Swaysland, *Familiar Wild Birds* (1901).

fails, the King, like the cuckoo that the Fool warned about in *King Lear*, cuts off Worcester's head.

Though he's usually thought of as an 'upstart crow', when considered at all in ornithological terms, Shakespeare himself may have been figured as a sparrow, according to scholar Helen Cooper. The play *Guy, Earl of Warwick*, published in 1661, but likely written in the 1590s as Shakespeare came to prominence, features a clown called Sparrow, born in Stratford-upon-Avon, who says of himself: 'Nay I have a fine finical name, I can tell ye, for my name is Sparrow; yet I am no house Sparrow, not no hedge Sparrow, nor no peaking Sparrow, nor no sneaking Sparrow, but I am a high mounting lofty minded Sparrow.'[11]

Hedge sparrow feeding a cuckoo, engraving.

Of all these Shakespearian sparrow references, the *Hamlet* quotation has had the most resonance. Variations on the line

The Fallen Sparrow, film poster, 1943.

'There's special providence in the fall of a sparrow' make their way into titles ranging from an article about Disney's portrayals of death, to a novel about a man whose daughter is killed by terrorists, to a book about terminally ill children. But its fatalism didn't take root in the same way as the comfort offered by the lines from the Gospel of Matthew. In 1904 Civilla Martin, wife of evangelist William Martin, was visiting an elderly couple in New York state who, despite their illness and frailty, found a

Envelope with a picture of a fallen Union soldier, referencing the biblical sparrow, 1861.

measure of happiness. After coming home, Martin wrote the lyrics to 'His Eye is on the Sparrow', inspired by the couple's confidence in the care of God.

The song goes, in part:

I sing because I'm happy
I sing because I'm free
His eye is on the sparrow
And I know He watches me.[12]

'His Eye is on the Sparrow' became a popular gospel hymn. Ethel Waters sang it at the White House and used it as the title of her autobiography. Along with 'We Shall Overcome', it was closely tied with the Civil Rights movement and was the favourite song of one of its leaders, Dr Martin Luther King Jr. The focus in the song is not on the importance of God's attention to all life or the scope of his panoramic view, but rather his care for a specific individual. The lingering phrase is 'he watches me'. The sense of foreboding in both the New Testament passage from Matthew and Hamlet's musing are stripped out. The fall is no longer central, leaving just a single singer, finding her own value, who is defiantly joyful, wild and alive.

The character of the not-quite-worthless bird also appears in India, China, and Japan. In his 1891 book *Beast and Man in India* John Lockwood Kipling, Rudyard Kipling's father, describes dishes of food and water set aside for sparrows near mosques and shrines. They are tolerated, though they steal grain right out of the market and interrupt prayers with their chirps. He writes, 'the sparrow seems to stand as the type of a thing of naught, an intrusive feathered fly to be brushed aside – but on no account to be starved or harmed.'[13]

Unlike European painters, who mostly ignored their resident sparrows or painted them dead into still-lifes (with the exception of Guercino), Chinese and Japanese painters incorporated the Eurasian tree sparrow and all its fluttering into their art. In the autumn, tree sparrows, like house sparrows, gather in flocks to pick at grain, the young of the year as well as older males and females, gleaning in groups and dodging the rain before settling down for the winter. The sparrow in the rice or bamboo was a traditional motif in the Chinese genre of 'bird and flower' painting developed during the eighth century. A fourteenth-century vertical scroll by Wang Yuan shows a male sparrow fluttering near the top of a bamboo stalk, a worm in its beak. Bamboo shoots tower over a rosebush with thorn-spiked stems. On the ground, near a rocky stream, a fledgling begs, wings spread; a male tree sparrow offers an insect for its open mouth. These naturalistic details – the thorns, the worms, the young bird's posture – characterized the style of this kind of painting. The sparrows' appearance in the grain was part of the natural cycle, signalling harvest time in the same way the stork and pine marked the start of the year and a cuckoo and iris might symbolize late spring.[14]

In Japan the tree sparrow, or *suzume,* was also a favourite subject of artists. The genre of 'bird and flower' paintings spread from China and the bird could be found in traditional Japanese images of autumn. But it also played a role in ink painting, which evolved in tandem with Japanese Zen Buddhism during the twelfth and thirteenth centuries. Monk artists sought out the commonplace as subject-matter, focusing on ordinary plants and animals, and often a few brushstrokes serve to capture a bird in flight or a wind-blown tree. The well-chosen lines catch the essence rather than worrying at the details. In a fourteenth-century painting by Ka'o, a solitary sparrow stands on one leg at the edge of a rock, staring up at fronds of bamboo, mouth

Yu Feng, Huang Zhou, Huang Miaozhi, *Birds in Snow,* painted in colours on paper.

Keisai Eisen, *Sparrow, Bamboo and Falling Snow*, *c.* late 1820s, colour woodblock print.

gaping. It is dwarfed by blank space extending over the edge of the cliff and reaching up into the sky. In a hanging scroll by the Buddhist priest Shokado Shojo (1584–1658), two fat sparrows weigh down a bamboo frond. They are roosting, but alert, ready to flick away on sharp wings. On the painting is a poem that reads:

> Among the bamboo lucid and green,
> Full of worldly dust,
> Sojourning sparrows lodge.
> In the silent woods, lingering fragrances
> And the murmur of human voices,
> There is no one.[15]

Shokado Shojo, *Bamboo and Sparrows*, hanging scroll, *c.* 1638, ink on paper.

Yuka, *Sparrows on a New Year's Mortar*, c. late 18th or early 19th century, woodblock print.

The birds' inconsequential nature worked its way into another religious tradition. Felice Fischer, curator of Japanese art at the Philadelphia Museum of Art, comments that the sparrows' fluttering appealed to the Buddhists: 'perhaps because they admired the free and spontaneous movement of the birds and drew an analogy to the enlightened man'.[16]

Sparrows could be purchased in China as food, but also as part of the Chinese Buddhist tradition of *fang sheng*, or the releasing of animals. Buyers demonstrated compassion by freeing a captive animal destined to be eaten, granting it life and freedom. The death is not foretold, and humans have the power to grant a reprieve. Monasteries often had fish ponds designated for *fang sheng*. The liberation might bring good fortune, or a bird could bear a prayer off on its wings. The sparrow was not the only bird used in this way, but certainly was one of the options. One official in the early 1600s describes going to the temple to commemorate his father, buying a sparrow (they must have been available just for this purpose) and setting it free.[17]

But in twentieth-century China, the gathering of the sparrows in the fields ceased being viewed as natural and benign. Their lives no longer held lessons or luck. In the late 1950s, as part of his Great Leap Forward, designed to make China competitive industrially with Western nations, Mao declared war on the 'Four Pests': rats, mosquitoes, flies and sparrows. The government claimed that the birds ate too much grain, taking the food out of the mouths of the people. Nature was the enemy of progress, and China would fight back with its most potent weapon – its large population. Rather than wasting bullets or spreading poison, citizens would literally scare the birds to death, making so much noise that they would be unable to land. In response to government mandate, children made posters in class reading 'Eradicate sparrows' and 'Sparrows are mankind's number one enemy'.[18]

Women with baskets of sparrows outside a temple, Yangon, Myanmar. A sparrow will be released in exchange for a banknote.

They were given the day off school and supplied with firecrackers to frighten the birds from the branches. Entire villages turned out to tear down nests, break eggs and kill the young.

Sheldon Lou, an elementary school student in Beijing at the time of the 'Four Pests' campaign, wrote in his memoir about the excitement of climbing up on the roof, seeing citizens flood into the streets, yelling and banging pots. But he also had misgivings, particularly when a flailing bird fell into the crowd, and he realized he couldn't tell what kind it was. Sparrow? Nightingale? Swallow? He calls to his classmate, 'How can we make sure we kill only sparrows but not other birds?' The boy jokes, 'How about teaching them birds some Mandarin, so they know we are not after them but only sparrows?'[19] Moments later, the two boys find a collapsed pigeon, the pet of a friend, which dies in their hands.

For many of those who participated in the sparrow campaign, killing a sparrow felt fundamentally different from swatting a fly, though both species were condemned. Not long after, in

大家都来打麻雀

1959, a short story set during the Great Leap Forward appeared in the *New Yorker*. In 'The Sparrow Shall Fall' by Han Suyin, the narrator returns from Shanghai to Beijing after the death of her father.[20] She emerges from the plane into a landscape where humans struggle to control the natural world. A sandstorm swirls yellow dust around the car, but she sees hundreds of saplings planted as part of a drive to grow more trees. The friend who drives her from the airport comments that all his professors are away building a dam. She arrives in Beijing as the city is gearing up for a three-day war on sparrows.

On the first day, everyone comes out to do battle. Girls wave flags to stir the birds into the air while boys wield shotguns. Firecrackers scare them out of any hidden crevice. A truck with a loudspeaker drives through the neighbourhood, broadcasting instructions to keep the sparrows flying so they can't roost and rest. Newspapers posted throughout the city announce the tally of dead sparrows. The narrator joins in, inviting sparrow-hunters onto the roof of her father's house, ringing the bell on a scarecrow to chase the birds away.

By the third day, the narrator is tired of the fight. She feels the birds' exhaustion as her own: 'On the third shrieking, booming, death-dealing morning, beneath my eyelids the sparrow fled, its erratic wings beat in my ears, its heart tick-tacked like a metronome gone mad. Whether I opened or closed my eyes, it was the same.'[21] When the city finally grows quiet, the narrator becomes nauseous at the thought of all the dead birds, but defends the government when a friend suggests the killing was 'ridiculous' and a 'waste of time'.[22] She is emotionally drained by the war that, according to the newspapers, did away with 800,000 birds in the city.[23] On the other hand, she tells herself, logic dictates that human life is more valuable than that of other creatures. The story ends with an open question: 'Was

'Eliminate the Four Pests', Chinese propaganda poster, 1958.

Travelling exhibition of sparrow hunting (36,000 birds), 19 April 1958, China.

it true that one or the other had to be, that the sparrow must fall so that man might live?'[24]

It was a question that lingered. Lou's unease, charted in his memoir, is shared by the adults. After the fight, his neighbour cheerfully deep-fries sparrows in his wok, but his own family refuses, and his grandmother whispers, 'Sinners, sinners, sinners. They were guilty enough to kill birds. Eating dead birds is a new sin added to an old sin. No one's able to expiate their souls.'[25] Weeks later, Miss Liu, a tiger expert from the zoo, comes for tea and her anger at the ornithologists who supported the sparrow eradication spills over. While the grandmother tries to shush her, worried that someone will overhear her criticism of the government, Miss Liu says, 'Didn't they know – I mean those scientists – that birds are our natural guards against pests and protect our crops?'[26]

In the wake of the campaign, which killed millions of the little birds around the country, insect populations swelled. An American soldier described webbed trees in the cities raining caterpillars into soup bowls.[27] A locust plague engulfed the countryside. According to Judith Shapiro, author of *Mao's War Against Nature*, many Chinese interpreted the infestation as a result of the sparrow campaign.[28] They viewed the attack on the sparrows as upsetting the balance of nature. The fate of humans and the little birds appeared to be linked rather than in opposition. Bedbugs replaced sparrows on the list of pests.

Fifty years later, with images of sparrows plummeting from the sky as part of their national heritage, the Chinese mulled over what to choose as their the national bird. The government leaned toward the red-crowned crane, with its rarity, elegant shape and mythical associations with longevity. But in an online poll on www.Tianya.cn, the sparrow won decisively. Voters noted that the scrappy bird, often dismissed but full of spirit, was more representative of the populace than the aristocratic crane. One site visitor commented, 'Their lives are unknown and unsung, like those of most Chinese people.'[29]

In this reappraisal of birds and citizens, a large population, anonymity and perseverance become measures of worth, elevating virtues that are overlooked and undervalued in the marketplace.

3 Dead and Dirty Birds

In roughly 64 BC, a young man wrote a set of poems to his lover. By many accounts she was older, and married to a powerful Roman, but despite these complications, the poems themselves were deceptively simple, the verses playful, suggestive, almost casual in tone. They are considered the first of a certain kind of intimate love lyric. It was, however, a *menage à trois*.

The first of Gaius Catullus' poems is directed not to his mistress, whom he calls 'Lesbia', but to her pet sparrow. The poet looks on as the bird nestles in her lap, nips her finger, eases her sadness. He wishes he could play the role of quieting her mind.

In the second poem, the bird is dead. Now, what comes between him and his mistress is not her affection for the sparrow, but her grief. She loved the bird more than her eyes, he writes, and has made them swollen and red with crying. On her behalf, he mourns for the sparrow: 'Now it goes to the darkened pathway/ Out of which, they say, none comes back.'[1]

What is the nature of this third wheel? The bird, held so easily in the palm, is as insubstantial as an emotion, as fragile as the potential between two people, as restless as passion, as alive as a heartbeat. The added element could be a description of the fluttering feeling of love.

Excavated wall painting from a Roman villa, *Swallow and sparrow confronted,* painted plaster.

As early as the Renaissance, though, when Catullus' works were rediscovered, critics read the word '*Passer*', translated in the first line as 'sparrow', as a slang word for 'penis'.[2] This gives a varying, though not entirely emotionally altered, interpretation of the poem and the figure of the bird in the lap which then lies immobile. It has lead to explicit speculation in classical journals about exactly what the poet and Lesbia may have been up to. Many readers hate this interpretation. It leeches the sense of the actual bird, substituting sniggering for delight. The scholar H. D. Jocelyn urged that the poem not be 'molested'[3] by this reading, calling it 'an example of learned silliness among wits and men of letters'.[4] But it persists.

Catullus admired Sappho, the Greek poet who, after her plea to Aphrodite to ease her lovesick heart, described the goddess arriving in a chariot pulled by sparrows 'beating the mid-air/ Over the dark earth'.[5] In Sappho's telling, they are the servants of the goddess of love. The birds would continue to inhabit this charged middle region between heaven and earth, love and lust

Sir Lawrence Alma-Tadema, *Lesbia Weeping over a Sparrow*, 1866, oil on panel.

– now a pulse of lechery, now a flicker of affection. The duality develops as a significant strand in European poetry.

In *The Parliament of Fowles*, Chaucer describes a meeting of all the birds, giving each an identifying phrase. The swan sings his own death, the cuckoo is ever unkind, then comes 'the sparow, Venus sone'.[6] The sparrow is now even closer to the goddess than just serving as her diminutive draft horse. The bird is her offspring, a minute, winged cupid. In *The Canterbury Tales*, Chaucer ties the bird to a different side of love, writing of the Summoner, an angry man with skin trouble who makes his living taxing people for their sins: 'As hot he was, and lecherous, as a sparrow.'[7]

After Angelica Kauffman, *Catullus and Lesbia*, 1784, etching.

Shakespeare's most famous sparrow image is in *Hamlet*, but elsewhere he plays on its desire. In *The Tempest* the birds are linked to Venus, and in *Measure for Measure*, they are hot-blooded. The hypocritical Angelo plans to put a man to death for impregnating a woman out of wedlock while himself plotting to sleep with a nun. He is characterized as planning to 'unpeople the province' with his morality: 'sparrows must not build in his house-eaves, because they are lecherous'.[8]

Other poets continued the tradition of associating the bird with love and desire, aiming low rather than high. John Donne played off Aristotle's description of the brief lives of male sparrows, shortened by their constant bouts of sex. In his poem commemorating a St Valentine's Day marriage, Donne writes of 'The sparrow that neglects his life for love'.[9] There's something almost romantic in this heedlessness of the future because the passion is so strong, but the sparrow in Donne's 'Progresse of the Soule' is not romantic at all. In this poem about the soul's migration through many animal bodies, Donne describes a sex-crazed young sparrow, barely out of pin feathers. He mates with his mother, then seeks out any available hen, not caring if it is his niece or sister, or who had been with her before him. Again, all this sex costs him days of his life:

> . . . freely on his she friends
> He blood, and spirit, pith, and marrow spends,
> Ill steward of himself, himselfe in three years ends.[10]

No lofty minded sparrow, this.

The cherished pet sparrow, dead before its time, makes another starring appearance in an early sixteenth-century poem by John Skelton called 'Phyllyp Sparowe'. The woman here is not an older lover but rather a young girl, a nun-in-training named Jane. The bird is slain by Gyb the cat and his distraught owner takes 844 lines to express her feelings on the subject.

Phyllyp, when alive, is full of personality. He has a name, which sounds like his 'chirrup' call. He chases butterflies and eats all the black flies he can find. He bites Jane's lip. He likes to be against her naked skin. Here he resembles even more closely Lesbia's sparrow which spends so much time in her lap. Jane senses this closeness might be improper, as he goes under her skirt 'Flyckerynge with his wynges!',[11] though she insists he 'He dyd nothynge perde [indeed] / But syt vpon my kne'.[12]

The poem's airiness comes from the contrast of the insubstantial bird with the weight of Jane's lamentations and the

Antonio Tempesta (1555–1630), *Mountain Sparrow.*

extravagance of her sorrow. She mourns the bird in a formal way, offering the Vespers of the Office of the Dead, contrasting Latin phrases with childish earnestness. She prays his soul will be preserved from hell and cries for vengeance on the whole nation of cats. Jane is frustrated that she doesn't have the education to compose a suitable epitaph, though certainly this is more individual attention than the death of a single sparrow ever received before, even from Catullus.

Unlike the literature that swirled around the court and centres of power, in folk and fairy tales the sparrow is less of a lecher and more of a trickster. In Grimm's fairy tales, which are full of birds – ravens, larks, willow wrens, bitterns, hoopoes – the sparrow has a personality larger than its size. In 'The Dog and the

Triumphant sparrow on an axe in 'The Dog and the Sparrow', from *The Brothers Grimm* (1886).

Sparrow', a sparrow befriends a hungry sheepdog, dropping bits of meat from the butcher and pecking at buns in the bakery until they roll into reach of the dog's mouth. The two settle down for a nap, the bird on a tree, the dog in the road. When a man driving a wagon approaches, the sparrow warns him that if he runs over the dog, the bird will impoverish him. The man dismisses the sparrow's ability to harm, and runs the dog over.

From then on, the man is plagued. The sparrow spills the wine, blinds the horses, calls all his sparrow friends to devour the man's grain. He dodges the man's wildly swinging axe until the horses are dead, the house windows are broken, the furniture is chopped to pieces. Every time the man wails about the 'unfortunate fellow that I am', the sparrow replies: 'not unfortunate enough yet'.[13] Finally the man swallows the bird, which reaches up out of his throat to taunt him. The man's wife heaves the axe at the bird and kills her husband. The sparrow flits off.

Another folk tale, in a similar vein, comes from Japan. The tradition in Japan seems markedly different from that of Europe,

Ohara Koson (1877–1945), *Sparrows in the Rain*, woodblock print.

Kajikawa family, Five-case black lacquer box in wood with a scene from 'The Tongue-cut Sparrow', 19th century.

focusing on the sparrow's flight rather than its mating, its freedom and childlike exuberance rather than an adult lechery, but the bird in 'The Tongue-cut Sparrow' is a close relative of the one described by the Brothers Grimm. In the tale, an old man raises a sparrow as a pet. One day, when the bird eats some paste set aside for starching clothes, his wife gets angry, cuts out the sparrow's tongue and sets the bird free. The man, missing and pitying the sparrow, goes off to look for it. When he eventually finds the bird and its family, the sparrow treats the the old man as an honoured guest, celebrating their reunion with feasts that last for days. But finally the man says he needs to go back to his wife. As a parting gift, the sparrow lets him choose one of two covered baskets. The man, feeling his age, chooses the lighter one. When he gets home and opens it, he discovers heaps of gold and silver.

When the woman sees these riches, she decides to visit the sparrow. The sparrow treats her rudely and doesn't give her anything to eat. When she is ready to go, she hints that she would like something to remember him by, and he lets her choose one of the baskets. Being greedy, she chooses the heavier one, but when she gets home and opens it, 'hobgoblins and elves sprang out of it, and began to torment her'.[14] The man, meanwhile, adopts a son and becomes wealthy. The story ends on a variation of 'they lived happily ever after': 'What a happy old man!'[15]

The sparrow in these two stories is meekness underestimated. The bird is the old woman in rags who turns out to be a witch, a prince disguised as the beggar at the door. The man in the Grimm story is thoughtless, but the sparrow's vengeance is gargantuan. The revenge is almost more horrifying because it comes from an apparently insignificant source. The deliberate and unnecessary violence of the man in the cart and the woman with the paste seem so small in the scheme of things. If you can't get away with discounting a sparrow or being cruel to dog, what can you get away with? God is keeping some elaborate accounts.

The most recent iteration of the trickster sparrow is in the movie *Pirates of the Caribbean* (dir. Gore Verbinski, 2003). When we first meet Jack Sparrow, played by Johnny Depp, he is captaining a tiny boat rapidly filling with dirty water. His eyes are smeared with eyeliner, his hair is snarled with earrings and ribbons, his teeth glint gold, and he takes his hat off in respect as he passes the skeletons of three hung pirates. He steps on dock just as his ship disappears beneath the water.

Tattoo on Johnny Depp's arm, seen when attending the film premiere of *Pirates of the Caribbean: Dead Man's Chest*, London, 3 July 2006.

He's landed on Barbados, where the governor's daughter, skin almost as white as her dress, is named Elizabeth Swan. When she faints from a too-tight corset, he dives in the sea to rescue her, only to be captured by the military as soon as he surfaces. He's reluctant to reveal his identity, but the Commodore yanks up his prisoner's sleeve to reveal a tattoo and sneers 'Jack Sparrow'. ('Jack sparrow' is slang for a male sparrow used in the same way 'cock sparrow' might be.[16])

The bird on his arm is actually a swallow, swooping split-tailed over the ocean. But the pirate's character distinctly recalls *Passer domesticus*. He's petite with a sparrow's nest of hair. He loves Tortuga, a rowdy city of fighting, prostitutes, drinking and the sound of breaking glass. Urchin-like, he thrives with few resources: coming ashore broke and boatless, he quickly picks a pocket and commandeers a fancy ship. He prizes survival over dignity or honour. After a long duel with the hero, Will Turner, he finally pulls a gun. 'You cheated!', Will says. 'Pirate', Jack reminds him.

Jack is happy in his pirate patchwork, but in other stories sparrows don fancy clothes. While the tales about the run-over dog and the tongue-cut sparrow show that slighting the common can have dire consequences, these dressed-up sparrows reinforce the idea that commoners should not reach above their station.

In 'Two Neighbouring Families' by Hans Christian Andersen, sparrows are aggressively ordinary. The neighbours of the title are a mother sparrow and her young, living in a swallow's nest next to a farmhouse, and a nearby rosebush. The roses, unaware of their beauty, appreciate the sun, their reflection in the pond, the featherless nestlings. They think 'It is so pleasant to have such merry neighbors.'[17] The mother sparrow, on the other hand, is sick of the vapid plants next door and hopes that when the cottagers pull up the roses they'll plant some corn. She scorns what she calls 'the beautiful', represented by roses and the peacocks that strut around the manor house. She tells her children, 'They only want to be plucked a little, and then they would not look at all different from the rest of us.'[18]

One day the mother sparrow is caught in a horsehair trap. The boys who find her are disappointed – 'It's only a sparrow'[19] – but bring her to a friend who offers to make her beautiful. The man covers the little bird in gold dust, puts a scrap of red fabric on her head as a cockscomb and sets her free. Terrified, she flies back home, but is plucked by her children, who don't recognize her, seeing their gilded mother as the dreaded 'beautiful'. She dies in the branches of the rosebush.

Throughout the tale, the sparrows remain resolute philistines. In the next generation, their lack of beauty becomes a lack of culture. At the end of the story, the rosebush has been transplanted to the museum of Thorwaldsen, a sculptor, in Copenhagen. The sparrows visit and peck the ground, blind to the great art around them. It's interesting to note that while the Grimm brothers gathered existing fairy tales, Andersen made up his own, so this personal view of the sparrow comes from an artist who grew up in a Denmark slum. In his view, the sparrows are an unappreciative mob.

Another disguised sparrow appears in a story from Pakistan about a cock sparrow who takes a new young wife. The old wife goes to cry under a crow's nest, but rain releases the dye from some scraps used in the nest, and she becomes peacock coloured and glorious. In her effort to become similarly attractive, the young sparrow wife throws herself into a scalding dyer's vat. When the husband, in an attempt to rescue her, drops and kills her, he shares his sorrow with the tree, who tells the story to the buffalo, who passes it on until the chain of mourning goes all the way up to the king. Once again the sparrow adopts second-hand beauty with disastrous results (and once again this slight life proves to have value).[20]

A third and darker breed of sparrows appears in some folk traditions, where the sparrow is essentially the anti-robin. Both are small birds that tolerate humans, but the robin is marked out from the everyday by its red breast. He also flits near houses, but eschews the large gangs that sparrows prefer. Robins linger in winter when many other birds have migrated, a splash of cheery red in the snow. 'Robin' is a nickname for 'Robert', making the bird seem all the more individual, all the more friendly. Robins are not just physically close to people, as sparrows are. They are sympathetic.

Stories say the robin's breast is red because it was pricked by the crown of thorns when the bird went to comfort Christ. Alternatively, he 'has a drop of God's blood in his veins'.[21] In the ballad 'Babes in the Wood' (1695), two children are abandoned in the forest by their uncle and die there. The only one who takes notice is a robin:

No Burial these pretty Babes
Of any Man receives,
Till Robin-red-breast painfully
Did cover them with Leaves.

In contrast to the kind-hearted robin, the folk-tale sparrow is not just drab; it can be downright devilish, as mentioned. According to Russian legends, all the birds helped protect Jesus in the garden when the Romans came to arrest him, but the sparrows' chirping led them to his hiding place. The swallows flew off with the nails for the crucifixion; the sparrows brought them back. As Jesus hung dying, the sparrows cried, 'He is living',[22] so the torture could continue. They are opposed to anything tender or pitying, a ruthless gang rather than a glowing flock.

So it should come as no surprise that when the nursery rhyme asks 'Who killed cock robin?' the answer is 'I, said the sparrow, with my bow and arrow.'[23] The rhyme appeared in the earliest extant book of nursery rhymes to be published, *Tommy Thumb's Pretty Song Book* (1744). The book, printed by Mary Cooper, measures 7.6 × 4.4 cm (3 × 1¾ inches) – scarcely bigger than a thumb.

After the first verse, which has the sparrow's confession, the story continues:

Walter Crane, 'Cock Sparrow', illustration to a book of nursery rhymes and tunes, *Baby's Bouquet* (1878).

Who saw him die?
I, said the Fly,
with my little eye,
I saw him die.

Who caught his blood?
I, said the Fish,
with my little dish,
I caught his blood.

Who'll make the shroud?
I, said the Beetle,
with my thread and needle,
I'll make the shroud.[24]

The rhyme in *Tommy Thumb's Pretty Song Book* ends there, but later versions like 'Cock Robin, A pretty gilded toy for either girl or boy', published in 1770, include descriptions of the owl, dove, wren, thrush and many more birds and their vows of what they will do to assist with the burial, from singing psalms to carrying the coffin to serving as parson.[25] Since this infusion of feathers changes the focus of the poem, likely it was added later. A list of mourning birds, without fish and fly and beetle, are imagined by Jane in 'Phyllyp Sparowe' as grieving for her dead pet. This may be the source for the bird section, or it may have even older roots.

Aunt Louisa's Big Picture Series, *Cock Robin*, book cover (1875).

Cock Robin (1888), version published by McLoughlin Bros.

In 'Who Killed Cock Robin?' the sparrow is arrayed, not just against the robin, but against the whole community of birds. At the end of the poem,

> All the birds of the air
> fell a-sighing and a-sobbing,
> when they heard the bell toll
> For poor cock-robin.[26]

It is unclear if the sparrow is among them. Many woodcut illustrations show him looking gleeful and unrepentant as he grasps his bow in his foot.

This is a clear tale of murder, with a confession and witness, but no motive. This unexplained crime left later writers with the urge to make sense of it. In 'The Happy Courtship, Merry Marriage and Picnic Dinner, of Cock Robin and Jenny Wren', commissioned by John Harris in 1806 as a prequel to the popular 'Who Killed Cock Robin', all the birds help celebrate the

Cock Robin, 1844, printed with brown ink on muslin.

wedding. Suddenly Cuckoo bursts in and begins to dance roughly with Jenny. Sparrow goes to shoot Cuckoo and kills Robin by mistake.[27] The senseless crime easily mapped onto political intrigue and murder mysteries. Some have read it as a parable about the downfall of Robert Walpole (particularly as the nursery rhyme was first published two years later). More recently, crime fiction has adopted the plot. The question forms the heart of *The Bishop Murder Case* by S. S. Van Dine and the spy novel *Who Killed Cock Robin?* by Margaret Duffy.

Disney's 'Who Killed Cock Robin?', a 'Silly Symphony' cartoon of 1935, opens with Robin serenading a sultry Jenny Wren with his guitar and red vest. A sparrow-shaped shadow passes across a tree and draws a bow. An arrow strikes Robin and he tumbles to the ground. In front of the Old Crow Bar, an ambulance howls up and a brawl results, punctuated by blackbird cops with Irish accents bopping people on the head. The scene moves into a courtroom with an owl judge and parrot lawyer. Witnesses called to the stand include a blackbird, 'Legs Sparrow' (a fat bird with spats, a bowler hat and a striped shirt, who insists 'I ain't saying nothing'), and a ditzy cuckoo. Jenny Wren, Mae West with tail feathers, struts in and insists that 'Somebody ought to be hung', and the chorus, led by the judge, insists 'We're going to hang them all.' An arrow pierces the judge's hat and Cupid shows up, declaring 'I shot Robin.' Robin, it turns out, isn't dead, only lovestruck. He revives and embraces Jenny Wren.

Here the sparrow is tied to Venus once again. It's a light, bloodless revisioning which conflates the sparrow with Cupid, and has moved far from the nursery rhyme. There isn't much levity in the *Tommy Thumb* version, with the fish and the beetle. In the original, the robin could be read as Christ (with whom he is closely associated), particularly because of the concerns in the original verses about witnessing the death and catching the blood. One of the legends about the Holy Grail was that Joseph of Arimathea used it to collect Jesus's blood as it dripped from his body on the cross. The sparrow once again adopts his role in opposition to Christ, here not just aiding the Romans but doing the actual killing. This leaves us with the sparrow painted with very dark colours.

This view of the sparrow as a figure of evil undoubtedly had its roots in the flocks that could raid a field and strip it of grain, taking much-needed human food. At various times the birds have had a bounty on their heads, and in the eighteenth century Englishmen formed 'sparrow clubs' to shoot them. The Egyptian hieroglyph for 'bad' or 'evil' is a bird that looks remarkably like a sparrow. The shade of the portrayal probably depended on the reality of the sparrow as a threat.

Many strands of the sparrow story come together in one of William Blake's *Songs of Innocence*, 'The Blossom'. Notoriously hard to interpret, the poem goes:

Merry, Merry Sparrow!
Under leaves so green
A happy Blossom
Sees you, swift as arrow,
Seek your cradle narrow,
Near my Bosom.

Pretty, Pretty Robin!
Under leaves so green
A happy Blossom
Hears you sobbing, sobbing,
Pretty, Pretty Robin,
Near my Bosom.[28]

The Bloſsom.
Merry Merry Sparrow
Under leaves so green
A happy Bloſsom
Sees you swift as arrow
Seek your cradle narrow
Near my Bosom.
Pretty Pretty Robin
Under leaves so green
A happy Bloſsom
Hears you sobbing sobbing
Pretty Pretty Robin
Near my Bosom.

Blake puts the birds in direct contrast. The sparrow is fast and chipper; the robin is handsome and sad. The illustration has no birds or flowers at all, but winged cherubs frolicking in the leaves of a flame-coloured tree. Two kiss. One reads. Another, older, bends over in her green gown. She may be holding a child; she may be not. But who is the happy blossom? Is it narrating the poem, similar to how Jane narrates 'Phyllyp Sparowe' as her pet nests nearby? Or is it a character in the story, like the robin and sparrow?

William Blake, 'The Blossom', from *Songs of Innocence and Experience*, c. 1824, relief etching printed in orange-brown ink and hand-coloured with watercolour and gold.

Like the Catullus *passer* poems, this verse has been read as overtly sexual, with the arrow-like sparrow and the 'cradle narrow' interpreted as representing male and female parts. The sparrow is happily lusty; the robin is crying with regret. It's hard to avoid this interpretation, given the sparrow's literary ties to lechery. Alternate interpretations have the blossom not as a young virgin, but as a baby, held by its mother, who relates the story of two birds. The child is happy, and is learning lessons about sympathy.[29]

This view is given strength by Blake's poem 'Auguries of Innocence' (which begins, significantly: 'A Robin Redbreast in a cage/ Puts all Heaven in a rage.'[30]) Near the end, he writes:

Man was made for joy and woe;
And when this we rightly know.
Safely through the world we go.[31]

The two birds are flip sides of human experience: low and high, urges and aspirations. The pesty side of our nature is not going away. Acknowledging this is part of gaining experience, of becoming adult. As Catullus and the young Jane show, a sparrow next to the skin may be distracting, but we are sorry when it's gone.

4 The Sparrow War

In the mid-nineteenth century, the United States was craving sparrows. Though the country had savannah sparrows, golden-crowned sparrows and thick-billed fox sparrows, they weren't the ones that nested in the eaves, the ones that had charmed Blake with their cheer. Caterpillars were infesting city parks, dropping off shady trees to land on coats and shoes, consuming leaves down to the the vein. Native birds were not doing enough to combat their spread. House sparrows had long shown a willingness to live in crowed, noisy, dirty areas that other birds might shun, and they were cheap. In 1847, a writer for the Philadelphia paper *The North American* urged their introduction: 'A very small sum of money would probably secure the object, and besides protecting the trees from the depredations of worms, would be the means of introducing a new species of birds about our houses and gardens, which of itself would be of more value than the whole cost of importation.'[1] The editor added that the suggestion inspired 'dreams of the sparrows twittering under our windows'.[2]

'A Jubilee Sparrow', from *Tales of the Birds* by W. Warde Fowler (1894), engraving. Here the sparrow makes its home on a rooftop.

In 1850 the directors of the Brooklyn Institute imported eight pairs of house sparrows, tending them during the cold months until they could be set free in the spring. But the birds died soon after release. They tried again the next year. Director Nicolas

Pike, on his way to a job as consul-general in Portugal, took the $200 the group had raised, ordered 'a large lot of Sparrows and song birds' from Liverpool and sent them to Brooklyn on the *Europa*. Sparrows were freed at the Narrows between Brooklyn and Staten Island and in the Greenwood Cemetery chapel tower, but it seemed that they too would die, so they were recaptured and taken inside a house for the winter. Finally, when the weather turned warm in the spring of 1853, the sparrows fluttered off.[3]

This introduction, so hard won, triggered a vicious argument, pitting ornithologists, preachers and politicians against one another, a feud so heated it was called 'The Sparrow War'.[4] The battles were fought with paper, reams and reams of it. These snowy piles contained statistics about whether dissected sparrow stomachs were filled with caterpillars or grain; testimonials from sparrow fans in England; tales for children laced with anti-sparrow propaganda; newspaper articles on acts of sparrow villainy; patents on sparrow-killing devices; and letters to the editor, editorials and reports of sparrow sightings in *The Springfield Republican, The Worcester Spy, The Wisconsin State Journal* and the *Idaho Avalanche,* among others. Writers probed the bird's usefulness and examined its character. William Cullen Bryant and others wrote poems in praise of it, and writers for *Forest and Stream,* a precursor to the sporting magazine *Field and Stream,* answered with damning doggerel.

The fight brought a tangle of emotions to the fore, ideas about nativity and citizenship as much as ornithology. During the second half of the nineteenth century and into the early twentieth, 30 million immigrants came to the US. Many settled in cities and the term 'immigrant' became increasingly associated with poor neighbourhoods and overcrowded tenement houses.[5] In the early 1880s, the U.S. government passed a bill to prohibit

Chinese immigrants from ever becoming citizens and Chinese men were lynched and shot in the streets of San Francisco and Seattle. Anxieties about increasing urbanization and the influx of foreigners all came to rest on the sparrow's little feathered head. The bird was usually referred to as the 'English sparrow', rather than the 'house sparrow'.[6] One ornithologist called the birds 'foreign vulgarians'.[7] The sparrow, linked to urban ills, was deemed sexually obsessed in its desire to breed and at constant risk of 'overcrowding, the result of its rapid propagation'.[8] Mocking poems associated sparrows with European Communists[9] and called the bird's song 'More tuneful far than Chinese gong'.[10] In contrast, a laudatory poem by Bryant claimed that the sparrow belonged firmly to the Anglo-Saxon world.[11] During the second half of the nineteenth century, the same newspapers covered the Civil War and the sparrow war. Who or what counted as an American?

The sparrow wasn't the only European bird brought to the U.S. during the nineteenth century. 'Acclimatization' societies all over the country were shipping and releasing anything with wings that might be a benefit to the park or countryside. The goal of the Cincinnati Acclimatization Society – 'the introduction into this country, of all useful, insect eating European birds, as well as the best singers'[12] – shows the tangle of desire, with strands both practical and aesthetic. The acclimatization society in Portland, Oregon, focused on German birds to make its large German immigrant population feel at home. Starlings, skylarks, English blackbirds, nightingales – all were set free, but none with as much frequency, or as much success, as the sparrow.

The poet William Cullen Bryant wrote a poem in celebration after seeing house sparrows hopping around a friend's garden in the late 1850s. In his 'Old World Sparrow' he writes:

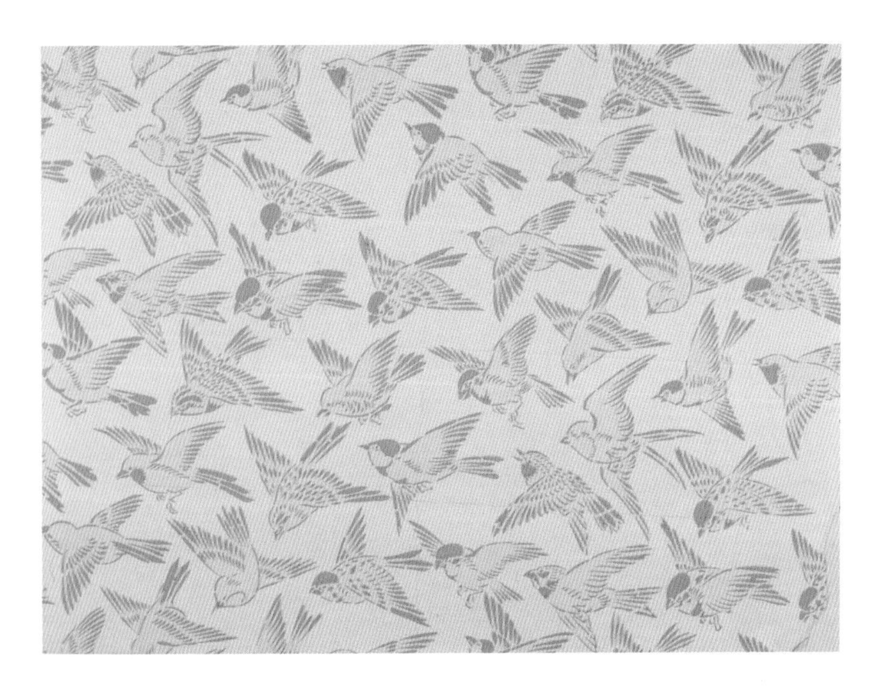

Thomas William Cutler, *Wallpaper, Print – Sparrow*, late 19th century.

We hear the note of a stranger bird,
That ne'er till now in our land was heard:
A winged settler has taken his place
With Teutons and men of the Celtic race.
He has followed their path to our hemisphere –
The Old-World sparrow at last is here.[13]

According to Bryant the sparrow is a specifically Anglo-Saxon warrior. He's a hard worker, with his 'busy beak'. The insect enemies to be vanquished by the sparrows are characterized by their foreignness: 'the pest of gardens – the little Turk', and 'Hessian fly'. They are a 'ravenous tribe', employing 'sly devices of cunning'.[14] Bryant is astute in noting that many of the insects plaguing crops during the nineteenth century were from other countries. They were primed for a population explosion in their new environment, just like the sparrow.[15]

Part of the attraction was a lingering doubt, left over from early New World explorers, about whether New World plants and animals were really as good as the Old World ones. This view was expressed most clearly by French natural historian Georges-Louis Leclerc, Comte de Buffon, in his theory of American degeneracy. He wrote,

> In America . . . animated Nature is weaker, less active, and more circumscribed in the variety of her productions; for we perceive, from the enumeration of the American animals, that the numbers of species is not only fewer, but that, in general, all the animals are much smaller than those of the Old Continent.[16]

Audubon and others were just starting to challenge this notion, with reports of spectacular American creatures like flamingos and bison.

For this among other reasons, city after city found the sparrows irresistible. Not long after they were freed in Brooklyn, sparrows were released in Boston and Rhode Island. Colonel William Rhodes claimed he brought them to Portland, Maine, in 1854, succeeding only on his third try. 'I imagine no living Yankee would wish to be now without the life and animation of the house sparrow in his great cities. They are like gas in a town – a sign of progress', Rhodes wrote to *Forest and Stream*.[17] He suggested bringing over blackbirds and starlings too. Philadelphia released a thousand sparrows. Sheboygan, Wisconsin, settled for six. New York, soon finding itself with a wealth of sparrows, began selling them to other cities. By 1875, sparrows

House sparrow in the city.

hopped around Rochester, New York; New Haven, Connecticut; Galveston, Texas; and Salt Lake City, Utah. By 1890, 33 states had imported sparrows.[18] The urgency to do so comes through in an article in *Colman's Rural World* from 1869. The author complains about street trees struggling in St Louis owing to coal smoke and insects. He cites the success of sparrows in New York and asks: 'Who will be the first to introduce a few pairs or a few dozens here, and place his name high on the list of public benefactors?'[19]

Communities banded together to ensure the sparrows flourished. Laws banned harming them. Bird-lovers built sparrow houses in their gardens and public parks and scattered grain for them during the winter months. A half-page illustration in an issue of *Harper's Weekly* of 1869 shows crowds in Union Square – men in top hats and bowlers, women in tiered skirts and bonnets, wonder-struck young girls – gathered around a bird mansion on a pole. The house is more than ten storeys high, and sparrows poke their heads out of the doors, hover above the roof and peck at the grass below, speckling the page like dark confetti. The caption to the illustration, titled 'The Sparrows' Home', expresses tempered optimism: 'A few were imported, but they increased so rapidly that it is now becoming a question in some localities whether they are not likely to become a nuisance themselves. They certainly are preferable to the worms.'[20]

The U.S. wasn't alone in its sparrow hunger. In fact, the entire globe seemed in the throes of a house-sparrow fever. One man brought twenty cages full to Argentina in the 1870s in the hope that the birds would eat bothersome insects. Brazilians imported

Walter B. Barrows, 'Sparrow in the City', from *The English Sparrow (Passer domesticus) in North America* (1889).

them to eat mosquitoes. Ship after ship carried them to Australia and, as early as 1868, citizens of Melbourne complained about the birds' attacks on fruit. After twenty years of living with house sparrows, the government imposed bounties on them – sixpence a dozen for sparrow heads. New Zealand acclimatization societies released them in Auckland, Nelson and Canterbury throughout the 1860s. Someone, possibly a soldier, set them free on Mauritius, an island off the African coast. Some of the attempts failed – the birds didn't survive long on Greenland after a release in 1880 – but most of the time the sparrows settled right down and began to breed.[21]

Slowly, notes of doubt began to creep in. In 1867, Dr Charles Pickering gave a talk at the Boston Society of Natural History, warning of the evils of these introductions. He pointed out that in many places in Europe sparrows were considered a

Stanley Fox, 'The Sparrows' Home', *Harper's Weekly* (3 April 1869).

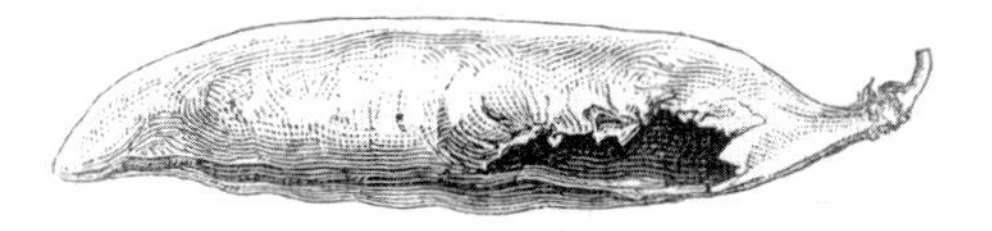

Peascod, emptied by a sparrow.

Wheat-ear—after the sparrow.

J. H. Gurney, two examples of house sparrow crop damage, from *The House Sparrow* (1885).

grain-devouring plague and had a bounty on their heads, and mentioned that French natural history writers had nothing good to say about them. They eat grain, not insects, he said, and quoted William Cowper to support his point:

> The sparrows peep and quit the sheltering eaves
> To seize the fair occasion: well they eye
> The scattered grain, and thievishly resolved
> To escape the impending famine . . .[22]

Were they really having an impact on caterpillars in the United States? Did farmers need to worry about their wheat? Was that 'chirrup' cheerful or annoying? As the debate got underway, the sparrows continued to conquer new territory. In 1869, as their fans flocked to Union Square, sixteen years after they hopped through the cemetery, sparrows had spread over about 11,000 square miles.[23]

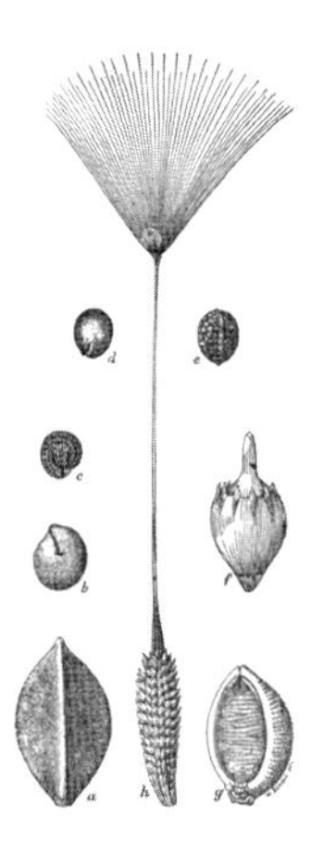

Sylvester Dwight Judd, 'Weed seeds commonly eaten by sparrows', from *The Relation of Sparrows to Agriculture* (1901).

The charge against the sparrows was soon taken up by Elliott Coues. Coues was only nine years old during the first sparrow introduction, but even then he had a passion for native birds. As sparrows were building their first nests in Portland, Maine, the 17-year-old Coues was taking notes about the migration of the eskimo curlew. During his early career as an army surgeon in the West, when sparrows were freed by the thousands, he discovered new kinds of warblers and vireos. And while others might not notice that bluebirds or purple martins no longer nested in a given spot after sparrows moved in, he did. And he didn't like it.

Physically, he didn't quite fit with Washington, DC, high society where he later found himself, with a prominent jaw he would eventually cover with a beard, a big straight nose and rough brown hair swept up off his forehead. There was something slightly raw about him, in his manners, in his words, in his bones. His fingers, adept at taxidermy (he could skin and stuff an English sparrow in one minute, forty seconds[24]), were tobacco-stained from chain-smoking. If he wasn't rolling cigarettes, he was writing, producing hundreds of books and articles over his lifetime on topics ranging from 'Nests and Eggs of the Clay-Colored Bunting' to 'Can Ghosts be Photographed?'[25]

In 1874, Coues wrote a short piece for *American Naturalist* reporting that sparrows were driving away robins and bluebirds. He noted, rather mildly: 'There is no occasion for them in this country; the good they do in destroying certain insects has been overrated.'[26] As he warmed to the fight, though, his language grew less circumspect. He savaged the birds, calling them 'wretched interlopers',[27] among many other names. His vehemence made sparrow advocates of those who might otherwise have stayed out of the fray, and he spared no language for them either. In one article he wrote, of one of the bird's champions, 'his name deserves to be stigmatized as long as there is a Sparrow left in the United States to shriek "Brewer! Brewer! Brewer!"'[28] In another, he called those who coddle sparrows 'a parcel of hysterical, slate-pencil-eating school-girls'.[29]

One of his more tactful essays, 'The Ineligibility of the House Sparrow in the United States' in *American Naturalist*, caught the attention of Henry Bergh, founder of the New York Society for the Prevention of Cruelty to Animals. With a high, pale forehead and drooping moustache, Bergh was tall and spectral, rather like a hovering conscience. His suggestions that beating horses be outlawed and real pigeons used in shooting matches be replaced with clay ones had met with scorn, especially in editorial cartoons. His long face made an easy caricature, but he built a receptive audience. He convinced New York to include all animals in its anti-cruelty law. A lecture in Cincinnati drew 1,200 listeners to hear him tell stories of dogs dissected alive and bulls bleeding in the ring, animals needing human sympathy that only the audience could provide.[30]

In his piece, Coues said he wouldn't go so far as exterminating all the sparrows, but did suggest that people stop feeding them, and that penalties for killing them be removed. Bergh responded with heat that matched Coues's own. In a letter to *Popular Science Monthly*, he called Coues an 'enemy to God' for his attack on 'that pretty little creation of the Almighty'.[31] The suggestion that boys be allowed to kill the birds was, according to Bergh, an education 'in the practice of murder', an apprenticeship after which 'they will be prepared to do the heavy business of throat-cutting, stabbing, and shooting'.[32] As a proponent of animal welfare, Bergh concerned himself with the rights and feelings of individual creatures, while Coues was more interested in the effect one species had on another, the roles they played in the larger ecosystem.

And, to those paying attention, there was little doubt that house sparrows pushed native species to the side. A 1889 report tallied 70 species that house sparrows harassed. They rousted purple martins from their homes, drove chickadees from backyard feeders and smashed house finch eggs. Migratory birds would return to their nests to find them occupied by house sparrows who defended them with beak and claw. Those who posted nesting boxes for bluebirds would often find house sparrows dive bombing the intended occupants, and sometimes killing them. A century later, one researcher in South Carolina discovered dead bluebirds in nestboxes with their skulls bashed in. She hid behind a blind and watched one house sparrow flit in and out of a nestbox with 12 bluebird nestlings in it, returning every few minutes while the bluebird parents were off looking for food. After an hour, all but one of the young bluebirds was dead, crowns bloody, skulls cracked. If that wasn't enough, she observed sparrows incorporating the corpses of dead bluebirds into their nests.[33]

Another person drawn into the fray was Henry Ward Beecher, a popular preacher at Plymouth Church in Brooklyn and younger brother of abolitionist Harriet Beecher Stowe. Beecher cut a dramatic figure, and not just because of his flamboyant lecture style. Long grey hair flowed past his ears; a blue cloak wrapped his large body. In the biopic, he'd be played by Russell Crowe. After a battering court case in which he was accused of sleeping with a close friend's wife (he was saved by a hung jury, but barely), he went back to the pulpit, where he advocated for evolution and women's suffrage. He had a lot on his mind.

But he took time out from railing against striking railroad workers to come to the little birds' defence. His essay in the *Christian Union,* subtitled 'Sparrows to the Rescue', starts out upbeat, declaring, almost mocking the subject-matter: 'We stand up for the sparrows!'[34] He writes that he enjoys watching them in his backyard, reports that Brooklyn remains full of native birds, commends sparrows' good work with the insects. Beecher criticizes bounties on sparrows and robins and crows. His final image is of birds nipping Coues 'until the last thread of his garment and the last hair of his head shall be borne away in triumph to line the nest in which a valiant sparrow shall give to its now native country another brood of these vigorous workers'.[35] As in Bryant's poem, the sparrows are immigrants: good, hard-working ones that any city would be proud to shelter. And Coues? Unlike Cock Robin, whose corpse the birds all volunteer to serve, in Beecher's vision Coues's death will be unmourned, even by the tender-hearted dove.

By 1878, the same year that Beecher and Bergh lambasted Coues for his cruelty, English sparrows in the USA inhabited about 640,000 square miles.[36]

Underneath the wash of Coues's bile lay the bedrock of reason. In 'The Ineligibility of the European House Sparrow in America', he wrote:

> In Europe these birds are part and parcel of the natural fauna of the country. They are not, as I understand, petted, pampered and sedulously protected from their natural enemies as they are here . . . [They] have their natural checks, and their increase is naturally checked.[37]

Released from the pressure of predators, populations could explode, as the English sparrow was proving as it conquered another town, another state. The lack of enemies gave the bird a competitive advantage over native birds and it was clear to Coues that bluebirds and martins were being driven out. He also complained about those who compared the sparrow in the U.S. to the sparrow in Europe. It is useless to look at how a species behaves in its natural environment as a map for its actions in a new landscape; the two situations are completely different, he explained. These are the same reasons that scientists today give for getting rid of non-native species. The arguments Coues laid out (and Bergh's opposition) would frame the debate about exotics for more than a century.

Even though they might not have evolved to prey on English sparrows, North American predators were starting to take notice. Owls flocked to Central Park. Northern shrikes, also called 'butcher birds' because of their habit of catching prey and sticking it on some convenient thorn to eat later,[38] began an English sparrow feast. The shrike, a grey bird with a black stripe over its eyes, looks like a bandit, and its efficient yet gruesome techniques make it easy to demonize. Professor Samuel Lockwood noticed as early as 1873 that shrikes were feeding on the newcomers, commenting on 'the ease in which a shrike in its winter visit gibbeted a sparrow in the city by putting its neck in the crotch of a small branch of a larch, and then, having knocked in the top of his head, the bird extracted it's [sic] victim's brains'.[39] A *New York World* reporter recorded a battle between shrike and sparrow in Madison Square Park in which the smaller bird fought hard but was 'torn limb from limb'.[40] A woman in Springfield wrote to the newspaper that shrikes were picking off the sparrows she fed, her pet flock.[41] In Boston, city gunners shot the shrikes picnicking on sparrows in the park.

While Coues laid out the scientific argument, the natural history writer Olive Thorne Miller presented the case for the sparrow as an undesirable citizen. The sparrows' habits were undeniably suspect. They brawled on the sidewalks. They bathed in dirt. They gorged themselves on grain from horse dung in the street. They had sex all the time, in public. In her article 'A Ruffian in Feathers' for *Atlantic Monthly* of April 1885, she distilled all the objections. In her presentation, the bird's music itself is an encapsulation of urban life. It 'harmonizes perfectly with the jarring sounds of man's contriving: the clatter of

'Tree Sparrow building its nest/Two Cock House-sparrows Courting a Hen' from *British Birds* by F. B. Kirkman and F.C.R. Jourdain (1950).

iron-shod wheels over city pavements, the war-whoop of the ferocious milkman, the unearthly cries of the venders, and above all the junk-man's pandemonium of "bells jangled out of tune."'[42] To Miller's ear, it's the only bird with nothing pleasant in its song.

In Britain, the sparrow had already become synonymous with urban life. As London grew grimier and more polluted in the early nineteenth century and other birds fled, the sparrow became linked to landscapes where only it could thrive. To some this city bird was a source of dismay, a commentary on London's disease, a devolution. It was associated with smog and chimneys, and many writers made a distinction between the town sparrow and its less wily, less grimy country cousin. In an article published in *The New Monthly Magazine* in 1869, 'A Chirp about Sparrows', the author looks at the filthy bird in the alleys and asks if it's even the same species lauded by the ancients:

> Degenerate, indeed, is the Sparrow of the streets if he be identical in race with the Passer of the classics. He is now recognised mainly as a Cockney chatterer, a twittering nuisance, a type of dingy impudence and pert familiarity and perky conceit. Can this be the pet birdie that was sacred to Venus, and to Lesbia so dear?[43]

To others, there was valour in its perseverance. The 'cockney sparrow' became a nickname for a kind of brash commoner, and eventually evolved into a stock music-hall character.

New developments in science allowed for an 'objective' view of the birds, and its character flaws seemed fundamental. James Hinton, a doctor and writer, looking at skeletons of sparrows and mice kept in glass cases in a bachelor friend's apartment in London, mused that the mice looked like miniature tigers.

> And as for the sparrows, their stuck-up, self-satisfied appearance, the pert and knowing look they put on, when thus reduced to their rudiments, surpasses imagination.

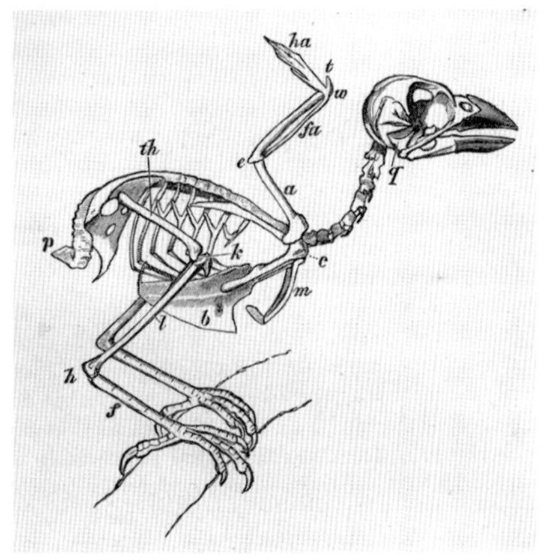

Arabella Buckley, sparrow in a branch and sparrow skeleton, from *Winners in Life's Race; or, The Backboned Family* (1901).

> The essence of the moral qualities of the bird seems almost to be concentrated in its bones. One can see that with such a foundation they could not be anything but what they are.[44]

Beyond her aesthetic objections in the *Atlantic Monthly* article, Miller charges the sparrow with criminal tendencies. It delights in a mob, steals crumbs from polar bears at the zoo and grubs from the hard-working robins who dug them up. Male sparrows here are not very progressive, but are rather chauvinist bullies. They change females on a whim, shriek down every objection and get their way, even on decisions about nest construction. A cock who lost a leg was able to hold onto his mate, though she, with a gentle prod, tried to get rid of him. Sparrows are slovenly housekeepers, who make nests of 'rubbish of all sorts and colors, from hay of the street to carpet ravelings from the spring house-cleaning'.[45] They're trashy and lower-class.

Miller tells the story of a hatchling in a spruce who gets tangled in the nest and hangs by one foot, unable to pull himself back in. The father tries to help, but eventually gets frustrated, wrenches his child free of the sticks and throws him on the ground. He then turns his back on his mate and the next day returns with a different female, one that seems younger and less tattered, to start a new family. As the author sums it up: 'It had taken that disreputable sparrow less than thirty-six hours to kill his baby, divorce his wife, and woo and bring home a bride!'[46]

At this point, it seemed unbelievable to most Americans that anyone had paid good money for sparrows. By 1881, readers were writing in to *Forest and Stream* to ask for techniques to get rid of the birds without poisoning their dogs in the process. The editor replies bluntly: 'Shoot the birds. That has been found effective in other instances. It will take some time and ammunition, but it is the only thing to be done.'[47] The Philadelphia Zoological Garden was catching them and feeding them to snakes. Readers mulled over the example of Vienna, where the city employed a sparrow-hunter to go after the ones in city parks. They swapped schemes in articles titled 'Hints on Sparrow Destruction'[48] and 'How to Kill the English Sparrow'.[49] One suggested hiring bored boys and equipping them with shotguns. Another wanted to requisition a steam fire engine to blast the nests with water and drown the occupants.

The magazine reprinted Bryant's poem in order to mock its hopefulness and naivety. Staff writer Fred Mather wrote a poem in response called 'The Old World Nuisance':

BANISH SPARROWS

The famous Dodson Sparrow Trap catches as many as 75 to 100 a day. Successfully used all over America.

Get rid of sparrows; native birds will return to your gardens. Sparrows are most easily trapped in July and August—young birds being most plentiful and bold.

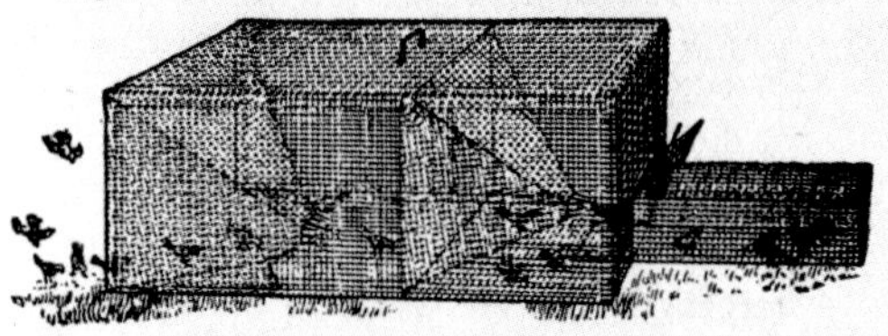

THE DODSON SPARROW TRAP

Strong wire, electrically welded. Adjustable needle points at mouths of two funnels. Price $5 f. o. b. Chicago

NOTE—Mr. Dodson, a Director of the Illinois Audubon Society, has been building houses for native birds for 19 years. He builds 20 kinds of houses, shelters, feeding stations, etc., all for birds—all proven by years of success.

Free booklet — tells how to win native birds to your gardens. Write to

JOSEPH H. DODSON

703 Security Building Chicago, Ill.

Early model of a sparrow trap, 1915.

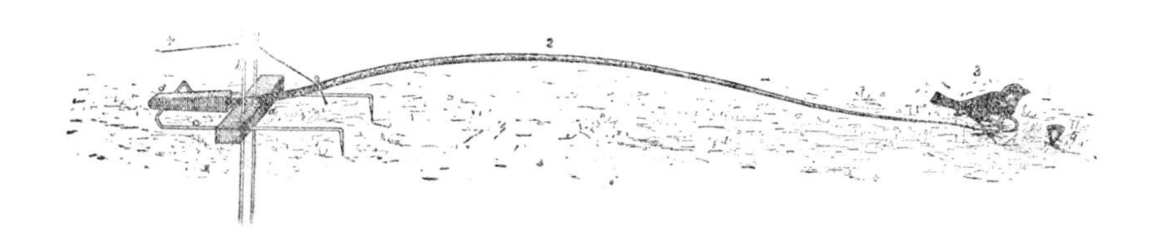

Walter B. Barrows, 'Sparrow Trap', from *The English Sparrow (Passer domesticus) in North America* (1889).

He has come to our shores with the murderous gang
Of Nihilists, who emigrated sooner than hang;
Paupers and regicides, red in the hand,
Which Europe continually sends to our land.[50]

He hones in specifically on the bird as immigrant of a different stripe than the noble Celts of Bryant's verse. Mather has in mind members of the anarchist and socialist movements that were building in Europe. Months before the poem was published a group called the 'Nihilists' assassinated the Russian Tsar Alexander II. After the failure of the socialist Paris Commune, which governed Paris for two months in 1871, many of the leaders immigrated to the U.S. In addition to comparing the birds to these radical Europeans, Mather cites Dr Coues, complains that the bird scorns worms and claims the bird has given a disease to the canary. Mather concludes:

American freedom has been much abused;
A home for the homeless we've never refused,
And the poor honest man can here cast his lot,
Bring his wife and his babes, and build him a cot;
But our long suffering people some morning will see,
Communists and sparrows thrown into the sea.[51]

Readers sent in their own contributions. One centres on the 'note of the stranger bird' praised by Bryant, describing it: 'The

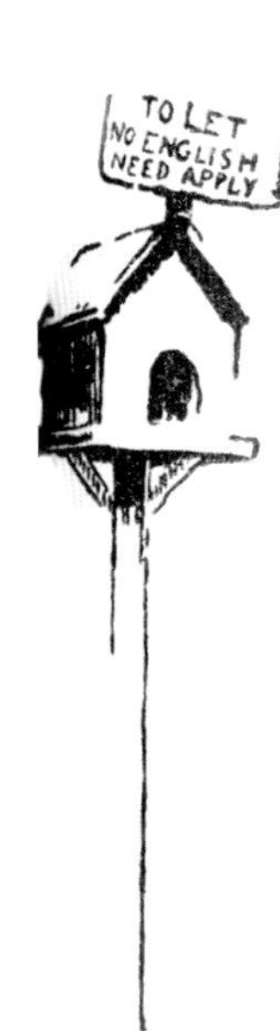

Mocking the anti-immigrant signs 'No Irish Need Apply', Ernest Thompson Seton sketches a bird-house that doesn't welcome English sparrows in *Lives of the Hunted* (1901).

entranced ear it does beguile/like rusty saw when scraped by file.'[52] A fantasy English sparrow obituary detailed the joy of the native birds now that the invader was gone.

Eventually, the sparrow was so far out of favour that natural history writer Ernest Thompson Seton included it in his collection of animal stories *Lives of the Hunted*, where the little bird took its place among bighorn sheep, black bears, coyotes and other creatures that might find themselves on the wrong end of public opinion and a shotgun.

Seton's hero is a bird named Randy, who in a way is the ideal sparrow, designed to overcome all objections to his species. He was raised with canaries in a basket, so he sings beautifully and builds a tidy nest. At least, he tries to, but his spouse, Biddy, keeps subverting his plans, throwing out the sticks he likes and adding a softer lining. First they build in a birdhouse set out by children, but abandon it after the inquisitive narrator slips a marble among their eggs. Their second nest is trashed by a man changing the electric lamp bulb over Madison Square. 'A Robin or a Swallow might have felt this a crushing blow, but there is no limit to a Sparrow's energy or hopefulness',[53] Seton writes.

Their domestic dreams are finally completely undone as Biddy gets tangled in a horse hair she's brought to their third nest, built in a tree in Madison Square Park, and hangs herself. In Seton's sketch in the margins, she dangles pathetically, wings drooping, one stick leg at an unnatural angle. The sparrow's bounce and twitter rarely give rise to tragedy, but here she seems like one of Dickens's urchins, coming to a bad end.

With a gold embossed bear on the cover, drawings of a sparrow choosing a strand of ribbon or chortling a canary tune next to the story, Seton's books are as sweet and tempting as candy. He is a champion of the individual creature, like the animal welfare groups. His characters have names and unique personalities,

Biddy hangs in a horsehair noose, in Ernest Thompson Seton's *Lives of the Hunted* (1901).

which make them easy to root for. But the author is also well versed in the science of the day. Biddy is notable for her white wing feathers, a trait written up in *Scientific American*, which suggested that 10 per cent of female house sparrows in New York had some white feathers, though this was a rarity in Europe.[54] Newspapers reported that sparrows had, indeed, been taught to sing like canaries. The whole piece, in a way, is a response to 'Ruffian in Feathers', where the male sparrows are brutish dictators. Seton's Randy is a sensitive artist and rather hen-pecked as Biddy tosses his cherished sticks to the ground. When Randy is brawling with a big sparrow named 'Bully', she joins in, deals out vicious pecks, and chases him off. Randy's privileged upbringing in a canary cage makes it tough for him to return to the gutter. He's sort of a Pygmalion in feathers.

The final battle of the sparrow war took place in Boston Common in 1899. Boston had been one of the sparrow's greatest supporters. After one importation from Germany in 1868, when only 20 of the original 200 survived, residents grew protective.[55] Additional introductions gleaned a few survivors, which eventually bred themselves into the hundreds. The chief forester of the city, charged with protecting the birds and feeding them in the winter, wrote that native birds like martins, robins and bluebirds were on the rise after the sparrow release. He also testified that the sparrows ate up canker worms and pests like yellow caterpillars.[56] But even here the sparrows were outlasting their welcome.

The city parks were overrun. A foreign tourist reported in 1880 that the Common was 'almost disfigured by the hideous miniatures of houses and cottages which are stuck up everywhere for the accommodation of this favoured representative of the old country'.[57] He was relieved to leave the city and escape 'Sparrowdom'.[58] A group of men organized the American Society of Bird Restorers as an antidote to the popular acclimatization societies. Their mission was to protect and encourage native birds and an explicit part of that was doing away with English sparrows 'without cruelty'[59] but with as much finality as they could manage. In 1889, the society presented a thirty-foot-long petition about the English sparrow to the mayor, requesting 'immediate reduction and suppression of this pest'.[60]

The mayor responded quickly, ordering five men to the city's public parks to get rid of the sparrows. Letters and articles poured in to his office, from those in England who declared the birds a nuisance to those who hailed them as scavengers, highlighting their work in city sanitation. Newspapers took sides,

Display for Sparrow's Empress Chocolates, manufactured by the Boston Confectionery Co. in Cambridge, Massachusetts, between 1870 and 1900.

with the *Boston Evening Transcript* defending the sparrows and the *Boston Journal* vilifying them. Rumors of cyanide and arsenic being mixed up by William Kennedy, chief of the Extermination Division, and plans to bring back shrikes to impale a few more sparrows fuelled the fervour. One woman promised to come to the Common and, as each nest was torn down, cry 'shame'.[61]

The mayor began to backtrack. 'I do not wish it understood, either by the friends or enemies of this bird, that I have taken any contract to banish it from the city limits, or even to materially reduce its numbers',[62] he announced. He promised to hold public hearings before setting out poison or traps.

On 13 March 1889, the sparrow team went out as planned, pushing a red cart, armed with poles to get at high nests. An audience trailed behind with 'expressions of horror on their faces, as though they expected to see horrible sights',[63] according to the *Boston Journal*. The team tore down nests built on the

WAR OPENS.

Raid Begun Upon the Sparrows.

Nests Torn Down on Common.

Mayor's Response to a Protest.

Foreman William J. Kennedy and his gang of six men started in on the sparrow hunt Monday at seven o'clock. They went to the deer park on the Common, and thence worked all across and through the park. At noon they had reached the Frog Pond.

The hoods of electric lights were found to be the favorite corner lot of the sparrow, and more than 30 of the cottages of the brown birds were piked out from these places. At the women's lavatory, every nook and corner sheltered a nest. Under the eaves the spaces left conveniently by the architect were filled, and more than 100 nests were had from this house.

Headline in the *Boston Journal*, detailing the efforts to remove house sparrows from Boston Common, 1889.

coverings of electric lights, dug them out of the eaves of the women's lavatory, scooped them out of holes in trees and stoppered them up so the birds couldn't rebuild. The headline of the *Boston Journal* article was 'WAR OPENS'.

After the purge, the mayor held a series of public meetings to determine if efforts should go further by killing sparrows and nestlings. A *Boston Journal* reporter noted, rather snidely, that many of the women wore hats that 'displayed the gay plumage of martyred birds'. He painted the scene as 'Sentiment' versus 'Statistics' and didn't seem fond of either.[64] The Society for the Prevention of Cruelty to Animals sent a representative to plead on the bird's behalf and detail its virtue as a street cleaner. The American Society of Bird Restorers sent ornithologists to describe the sparrows' ill effects. Finally in mid-April, after the second hearing, the mayor called off even the nest destruction, saying it was taking up too much of the time and resources of his staff.

The Sparrow War would reshape the way the country thought about non-native species. By the turn of the twentieth century, most scientists agreed that the haphazard introduction of new species was a bad idea. In the Lacey Act of 1900 the federal government banned the importation of specific species, the house sparrow among them. But as Coues, who died the previous year, noted, the real winner was the sparrow.

'I could whip all my featherless foes, but the Sparrows proved too many for me, by a large majority, and I retired from the unequal contest some years ago', he wrote.[65] By 1899, they had spread over 3 million square miles of America, inhabiting almost the entire country.[66]

A few years later, in 1906, a man in Pittsfield, Massachusetts, began to eye his canary suspiciously. He'd bought it at the Great Barrington Fair, but it didn't sing much and left traces of yellow in the bath. A few months later, in the small town of Rutledge, Pennsylvania, a man went door to door offering German canaries for $1, guaranteed to sing. When the little birds died within days, the owners discovered that underneath the promised melody and gold paint lurked an impostor: that common, worthless bird, the English sparrow.

5 The Fall of a Sparrow in Providence

On 31 January 1898, as the Sparrow War trickled to an end, a vicious blizzard slammed into New England. Heaps of snow stopped trains and streetcars in their tracks; automobiles and horse-drawn carriages floundered. Drifts reached the tops of fences. High winds tore at telegraph and telephone wires, isolating neighbourhoods.

The next day in Providence, Rhode Island, as utility workers made their way through a snarl of downed wires and poles snapped in half, Hermon Carey Bumpus found a flock of house sparrows at his feet as he walked uphill to work. A popular biology professor at Brown University, Bumpus was known for his good nature – letting students roast and eat their lobsters after studying them – and his rigorous field investigations. The birds usually spent the winter sheltered by the vines of the Providence Athenaeum, but the storm had been too much for them, and they lay distressed and dying on the sidewalk.

Even before the cold snap, Bumpus had been intrigued by the introduced sparrows. At the time the sparrows were first set free in Brooklyn, Charles Darwin had not yet published *On the Origin of Species*. He was still mulling over the nature of variation, including that found in the finches of the Galápagos Islands. These birds, glimpsed on his five-year voyage on the *Beagle*, had a wide array of beaks, each a shape and size ideal for exploiting a certain kind of food. The isolation of the islands and the scarcity of other bird life meant that finches occupied almost every available niche. How did that happen? In 1859, before house sparrows wore out their welcome in the U.S., Darwin laid out his argument for evolution by way of natural selection. It was a breathtaking theory, but inspired a pressing need for examples of natural selection at work shaping species, carving one from another. Looking at the faltering birds, Bumpus thought he might have just such an example.

Most of the imported birds were originally from England and Germany, so they didn't vary much before they hopped out of their cages. From this small population, they spread all over the country, tailoring themselves to niches from alpine meadows to grasslands to swamps, undoubtedly changing in the process. Bumpus had already compared house sparrow eggs from England to those in the U.S. and found the American ones lighter and more variable. The storm offered a chance to watch the birds respond to a crisis, rather than the slow pressure of time.

Back at the anatomical lab, Bumpus observed the storm-downed sparrows. In the hours and days after the blizzard, as the city righted itself, some of the sparrows lived. Others died. Bumpus measured all 136, noting body length and wing spread, head width and weight. Comparing the 64 corpses to the 72 survivors, he concluded that the storm culled the larger and smaller sparrows from the population, eliminating the extremes.[1] 'It is the type that nature favors',[2] he declared in his paper 'The Elimination of the Unfit as Illustrated by the Introduced Sparrow, *Passer domesticus*', noting that there were many more 'freaks'[3] among the dead than among the survivors. He read into these tiny skulls an example of the still controversial notion of 'the operation of natural selection, through the process of the elimination of the unfit'.[4]

Bumpus published all of his sparrow measurements in the original paper, and his data has been reinterpreted many times in the intervening years. Whether or not he was correct in exactly what the small deaths meant for the species, he was the first to use the birds to study the mechanisms of evolution. And Bumpus' early and pioneering experiment was only the start. Throughout the twentieth century and into the twenty-first, house sparrows remain an important tool in our understanding of the way selection works. As non-native invaders, in the U.S. they are considered 'trash birds'[5] without legal protection so are endlessly available to researchers to use however they like. They are close at hand, obtainable by leaning out the window or setting a trap on the lawn, so their study doesn't require a large travel budget. House sparrows are a poor man's Galápagos finch.

In 1917, zoologist Joseph Grinnell was stunned to find house sparrows living in palm trees and cottonwoods near the defunct mines of the Pacific Coast Borax Company in Death Valley, California, 54 metres (178 feet) below sea level. Death Valley, a stark landscape of sand dunes and salt-baked earth, is notoriously hot and dry. Some years, the annual rainfall is zero. The Valley hosts plants whose roots stretch many metres deep to find any water, and the kangaroo rat, which can live its whole life without a drink. Only four years earlier, in 1913, the temperature had reached 57°C (134°F) in the shade.[6]

The property manager told Grinnell that the sparrows arrived as the railroad was completed, 27 kilometres (17 miles) away. Maybe they rode in boxcars, or followed a trail of crumbs left by workers. They had already shown a disconcerting ability to track people wherever they went. Harsh ecosystems seemed to pose no obstacle. By the time Grinnell discovered them in Death Valley, house sparrows had found a home in all the states of the U.S. and

Hawaii. But staking a claim in country this forbidding was almost laughably bold. Was there any place where they wouldn't move in? They were mocking the limits of survival.

Grinnell, like Bumpus, saw the scientific potential. He termed the release and settling of house sparrows in these diverse environments 'an experiment in nature'.[7] It was an ideal opportunity to gauge how long it took for a subspecies to develop, to watch evolution in action. 'In nature, subspecies have differentiated under just the conditions self-imposed by the English sparrows through their powers of invasion',[8] Grinnell wrote in an article about the Death Valley flock. He took a few for his collection, though he didn't immediately see any deviations from those in the San Francisco Bay Area, where he ran the Museum of Vertebrate Ecology at the University of California at Berkeley.

Gradually, the experiment in nature bore fruit. Those who cared to look noted differences among populations of house sparrows. About 40 years later, biologists Richard Johnston and Robert Selander collected house sparrows from all over North America and the Hawaiian Islands. When the birds were placed side by side, the changes were clear. In Edmonton, house sparrows were big and bulky. In Mexico City, they were petite. In Vancouver, British Columbia, the birds were dark; in central Texas, far more pale. In Zachary, Louisiana, they had yellow underbellies. In Oahu, they had buff-coloured legs. The birds whose ancestors Grinnell saw in a Death Valley palm were tiny and exhibited 'extremes of pallor'.[9]

The scientists were surprised by the stark differences in size and colouring – local populations could almost be classified as subspecies – and how rapidly they occurred. Early on, evolution was thought to take thousands of years of tiny shifts. But Bumpus demonstrated that one storm could shape a population, and Selander and Johnson showed that less than a century produced significant alterations. Evolution appeared to be a matter of decades rather than millennia.

All these house sparrow measurements, including those made by Bumpus, have been used to look at how animals might be affected by climate change. Critters that live in warm areas are, as a general rule, smaller than similar ones that live in cold areas. Bigger bodies create and hold more heat, giving them an advantage in chilly climates. Johnson and Selander saw this borne out by the house sparrow study. The birds in snowy Canada were the largest; those in Central America were among the smallest. The size of a bird's body overall affects the size of its wings, since a certain length is needed to keep a given bird aloft.

With this in mind, William Monahan at the University of California looked at the length of sparrow wings over the course of several years in Honolulu, Berkeley, Detroit, Denver, Lawrence in Kansas and Los Angeles. He plotted these lengths against the coldest winter temperature of those years, and showed that as the temperature was growing warmer, the wings were growing shorter. At the end of his article, he wondered if, as global warming continues, the birds' overall size (which might not be governed by genetics in the same way the length of the wing bone might be) might be changing more quickly than the wings can evolve.[10]

Because they are so plentiful, house sparrows have been used in all kinds of scientific studies, not just those centred on evolution. In the seventeenth century, the dawn of science as we know it, Robert Boyle used sparrows in his experiments with a vacuum pump, showing they cannot breathe (or live) in a container from which all the air has been removed. Antonie van Leeuwenhoek examined the brain of a sparrow through his finely tuned microscope. An early experimenter with electricity determined that they lost weight when exposed to current.

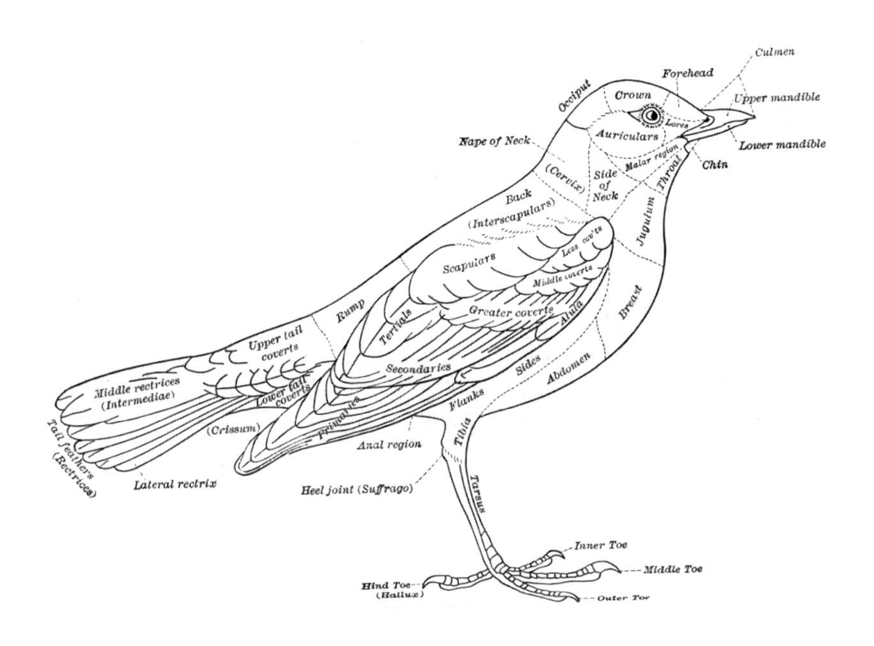

Vernon Lyman Kellogg, diagrammatic outline of a bird's body, from *Elementary Zoology* (1902).

More recently, house sparrows have been used to answer the question of what makes a non-native species a good invader. This is a pressing issue, as invasive species are changing many ecosystems, driving away native species and reducing biodiversity. After habitat loss, non-native species are the primary cause of extinctions, and improved transportation globally has made their introductions all the more frequent. The house sparrow is a logical choice for these investigations, as it is the Genghis Khan of the avian world. When house sparrows arrive, a number of other bird species invariably disappear. What did it mean that it could live everywhere?

One reason for the sparrow's success appears to be its flexibility in terms of behaviour, particularly when moving to a new place. One study compared sparrows new to Panama with those that had been in Princeton, New Jersey, for 150 years. The invading birds tried and ate new foods (yogurt mixed with dog food was particularly popular) much more quickly than those that had lived in one place for a long time. Willingness to sample a different seed or berry or build a nest in an unfamiliar tree (or streetlight) would be an undeniable advantage to a bird alighting in a foreign environment.[11]

The desire to hang out in big flocks, chattering and taking up the sidewalk, may also work to house sparrows' benefit. Scientists challenged sparrow flocks to figure out how to pry a lid off a seed container. Some groups had two birds, some had six. The result: bigger flocks took less time to get to the seeds. (City birds also did better than those from the country, so urban sparrows may have developed skills to help them survive in a landscape of frequently evolving potato-chip bags and soda cans.[12]) In this, they are like people, who also solve problems more quickly in groups.

A final reason for house sparrows' success is undoubtedly their toughness, evicting other birds from their nests, pecking at each other. While in some animals, bright colours or long tail feathers might indicate a male has energy and health to spare to grow such fancy ornaments, in house sparrows the size of the black badge on the male's chest indicates he is ready for battle. The more aggressive he is, challenging high-ranking males and fending off attacks on his own status, the larger a badge he will grow after the autumn's feather moult. Unlike a change in the genes, which might govern wing length, badge size is tied to phenotypic plasticity: it is governed by a change in the environment, in this case the bird's relationship with its peers. The size of the badge reliably predicts who will win a fight.[13]

All of these traits combine to make a very hardy bird. 'They are survivors', says researcher Denis Summers-Smith, who is known as the 'sparrow guru'[14] because he's spent so much time

contemplating the birds. 'They are extremely successful in a world that is dominated by human beings.'[15] This adaptation to the built human environment, more than anything else, enables the house sparrow to persist. Summers-Smith became interested in the natural world early on, when an uncle tutored him in botany. When he returned from fighting in the Second World War after being wounded in France, he moved to the south of England, and started his career as a chemical engineer. He was still interested in natural history and knew he wanted to do an in-depth study of one species, but he'd married and fuel was rationed, so he couldn't travel far. Christmas Island frigate birds and Australian bowerbirds were out of the question. And there the sparrows were, on the streetlights, on the park statues, by the railroad tracks. 'It was a common bird and no one was interested in studying it',[16] he says.

His interest eventually stretched to cover the entire *Passer* genus, and pushed him far beyond his usual habitat. He travelled to Kenya to look at the Swainson's sparrow, Israel to see the Dead Sea sparrow and India to observe the Sind Jungle sparrow. He's followed them up trees and down mines. After 60 years of studying sparrows, Summers-Smith still admires their perseverance. Now that he's in his nineties, he doesn't venture out as much, but keeps up his surveys near his home in Guisborough in Great Britain, looking at the birds that hop nearby.

Another sparrow that just happens to be often on hand is the white-crowned sparrow. The white-crowned sparrow's name describes three white stripes on its black head: one down the centre and one above each eye. The contrast between the deeply pigmented cap and the bright white makes the stripes almost glow. They don't live on roofs and storefronts like the house sparrow, but don't mind gardens and ornamental shrubs, so they are not hard to find. Of the five subspecies, four migrate. One subspecies, *Zonotrichia leucophrys nuttalli*, hangs out in coastal northern California all year round. The biologist Luis Baptista, who studied the white-crowned sparrow for many years, called the it 'the white rat of the ornithological world'.[17]

Baptista heard them singing near the law school, up the canyons by the botanical garden, and all over the campus of the University of California at Berkeley where he worked. Born in Hong Kong, Baptista grew up thinking about birds, music and language, and the relationships between them. As a boy, he raised canaries. He sang to them and, as they grew older, they sang his songs back to him. Questions about dialect, accent, creole, improvisation and bilingualism came easily to an immigrant who knew five languages and whose ancestors spoke a mixture of Portuguese and Cantonese.[18] He decided to use white-crowned sparrows to investigate a different kind of evolution – the cultural evolution of changes in song.

Baptista carried a tape recorder around the Bay Area, from the mist-soaked chaparral of the Presidio army post in San Francisco, to islands that dot the bay, into industrial Richmond, along remote paths in the Berkeley Hills. He captured songs from individuals, and sampled a number of birds in a given area. He also played songs back within sparrow earshot, watching as sometimes listening birds fled, and sometimes they attacked his machine.

Unlike the song sparrow, which can have up to 15 dramatically different songs, the white-crowned sparrow has a comparatively plain tune which it learns early in life, usually before it is two months old. To the untrained ear it sounds like two long whistles, rather like the pulling back of a slingshot, before the bird lets fly with a scatter of pebble-like notes. To the trained ear, like Baptista's (who could identify individual sparrows based on the way they sang), this pattern contains dozens of variations,

John James Audubon, 'White crested sparrow', now known as the white-crowned sparrow, from *Birds of America* (1827–30).

White-crowned sparrow.

telling him the bird's gender, family relationships and home turf. The sparrows pack a surprising amount of improvisation into their basic notes.

Back at the the lab, Baptista used spectrographs to compare white-crowned sparrow songs. These visual representations of songs can look like someone writing with a fountain pen on a bumpy bus: black scratches on a white page arranged according to frequency. Individual songs can look uncannily like notes arranged on a musical stave. By looking at the song pictures, Baptista could tell that though the overall outline of the two-second song was the same, populations sang, or 'spoke', different dialects. In Tilden Park, the white-crowned sparrows sang 'whistle, buzz, trill'. In other parts of Berkeley, they replaced the second buzz with a whistle. Some groups ended their songs with a fanfare of notes. Others kept it simple. Even within the same city, tunes could change between neighbourhoods. Baptista identified Presidio, Lake Merced and San Francisco dialects, the last sung by birds throughout the city. A small flock in Union Square, isolated by towering department stores and posh hotels, had its own version, one that ended with matching twin notes.[19]

Song sparrow.

Some birds were bi- or even trilingual, singing one song to one group, another to another.

But why all these fine gradations of noise? Males do the most of the singing – testosterone prompts song learning – and the tunes can be used to woo mates and define territory. The variations might just be one bird mishearing a tune and teaching it to his children, spreading it through the flock like a mutant bit of DNA. Perhaps some birds take the theme and improvise, white-capped jazz musicians. A distinct dialect might be a way for local girls and boys to find one other, advertising their adaptation to the conditions of a given district.

In an effort to find out the dialects' purpose, Baptista and biologist Martin Morton conducted a study of white-crowned sparrows in the high Sierra, near Tioga Pass, just east of Yosemite National Park. They tested to see whether young birds mostly learned the dialect of their fathers, who would sit and sing after bringing them their food. They didn't. Instead they seemed to be learning the songs of birds they would be most likely to fight ('fight' in this case meant flinging songs at each other), even if those birds were of a different species. The scientists documented white-crowned sparrows mimicking tunes of a Lincoln's sparrow and strawberry finch, birds they heard in their immediate environment and might have to sing down in a contest over territory. Peers, rather than parents, seemed to dictate the dialect.[20]

Morton had been following this alpine population of white-crowned sparrows (*Zonotrichia leucophrys oriantha*) for years, examining it from almost every angle. He studied the way they migrate the moment they store up enough fat, and the way their bills grow longer in mid-summer when there are more insects to be gleaned, then shrink back down in the autumn to eat seeds. He looked at how often members of a mated pair would cheat on each other. (Very often, it turns out. More than 30 per cent of young were unrelated to one of the birds tending the nest.[21]) After 40 years of monitoring, these alpine sparrows are still under the microscope, making this one of the most intensely studied populations of wild birds.

Tioga Meadow, 2,743 metres (9,000 feet) above sea level, is laced with streams. In late June, the sound of water is constant – dripping, trickling, roaring under patches of snow that linger on mountain slopes and crust over the grass. Unlike many house sparrow haunts, it appears untouched, but a closer look reveals intense human documentation. Nests in willows and stumpy pines are tagged with flagging (not right by, which might alert predators, but close enough for researchers following clues in their notes to be able to find them again), birds are marked with an individual sequence of bands, and family trees are mapped back for generations. Some of the sparrows have tiny radio transmitters that loop around their legs, long antennae trailing down their tails.

It's a precarious existence. Unlike the white-crowned sparrows in San Francisco, which don't migrate, these white-crowned sparrows come to Tioga Meadow only in the summer, arriving from Mexico while the meadow is still buried in snow. In the first few weeks they live on butterflies and ladybugs swept out of the Central Valley by storms and dropped in the mountains. Birds claim patches of meadow as it thaws, chasing off intruders and starting to build nests. Blizzards sweep in, causing some to flee their territories, and others to tough it out in the cold. Ground squirrels and Clark's nutcrackers are hungry and on the prowl.

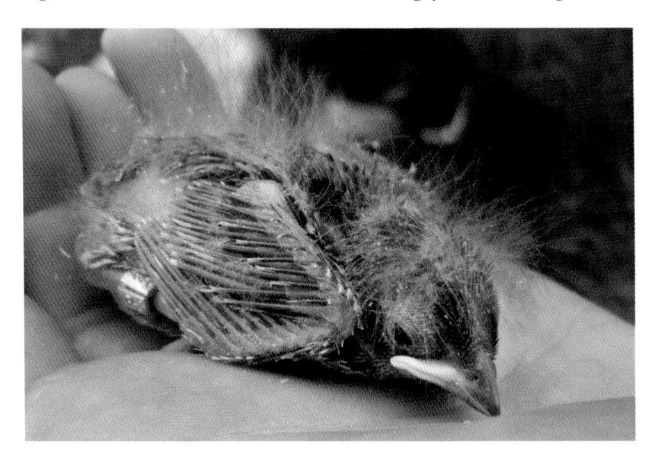

Nestling mountain white-crowned sparrow.

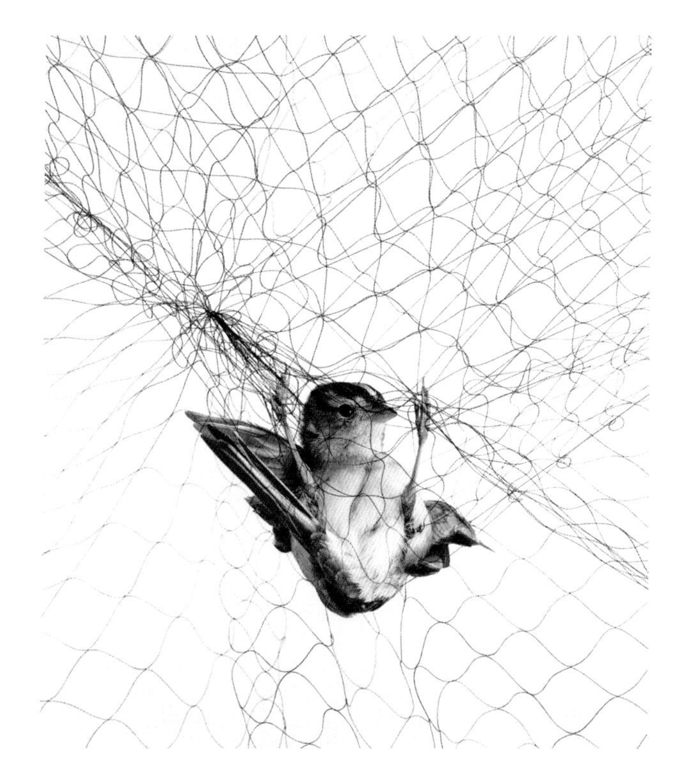

Todd R. Forsgren, *White-crowned sparrow (Zonotrichia leucophrys)*, 2006, photograph.

Now, in early summer, the sparrows are laying eggs and defending their nests. Willows are turning yellow at the tips, just cracking open into fluff. Researcher Andrea Crino is stepping over the raised burrows of ground squirrels, eyeing the willows. She is searching for white-crowned sparrow nests, one of her favourite things to do. 'You get strangely addicted to looking for nests, particularly once you find one', she says. 'It's like an Easter egg hunt with real eggs.'[22]

Crino is a graduate student in Creagh Breuner's lab at the University of Montana, where biologists study not songs, but stress. In her trailer, she has a copy of Morton's book, the bible of this extended sparrow family, but there's more to discover. The researchers are looking at, among other things, the trade-off in males between the speed of the reaction to stress and the development of macho traits like a vivid crown or aggression. 'If you're a male with a big stress response you're less attractive to females, but you might survive longer', says Crino.[23] On the other hand, if you don't get stressed as quickly, you might be more sexy, but only last a year, she adds.

Stress triggers the release of a specific hormone into the blood, and the researchers gauge the bird's stress response by testing how quickly and in what amounts the hormone is present. They are also looking at stress in females, and whether they pass on their stress responses to their offspring. Researchers catch the birds in seed-laden traps, and take blood samples to

Recently hatched mountain white-crowned sparrows.

Fledgling mountain white-crowned sparrow.

test the levels of stress hormone. In both sexes, they look at how stress levels determine whether the sparrow stays or flees when a storm hits. The researchers want to find out how these decisions translate into many chicks or a long life.

Crino checks on marked nests, counting and measuring eggs, taking blood samples from nestlings, keeping her eye open for undiscovered nests. She ignores the few taller or more established trees, as do the birds. 'They seem to like really crappy vegetation', she says. 'Who knows why.'[24] One nest by a tiny stream has been abandoned, though four cold eggs remain – jade with rust speckles – along with a few red pine needles. Finally, researcher Brett Klaassen van Oorschot calls over: 'Hey, I found a nest.'[25] Right on the ground at the base of a shrubby pine, it's woven into the grass at the edge of a muddy trickle of water. 'This is a good sized nest', says Crino, anticipating many eggs, a whole new generation of stress data.[26]

Meanwhile, house sparrows live in the town of Lee Vining several miles away, where it is warmer in the winter and they can make an easy living off tourist scraps. Even if they were buffeted by storms, they might not mind. Breuner tells me that the house sparrow is unique in that it shakes off trauma like water. Unlike the white-crowned sparrow, if you capture a house sparrow and put it in a cage and test the level of its stress hormones the next day, they will be back down to normal as if nothing had happened, she says. A downtown traffic jam, the rumbling of large machinery, life in a cage: they are impervious to stress.

These studies are useful because they push us beyond our limited human perceptions, opening a window into the lives of wild animals. But they also tell us about ourselves. How does a southern drawl or clipped northern vowels affect our views of the speaker? What purpose do they serve? Or, for that matter, what the purpose of music? Is it better to be cautious than sexy? How do our physical environments change us? For instance, as we become more urban (researchers suggest that in about 2007, the balance tipped so that, for the first time, more people worldwide lived in cities than in the country[27]), how will we respond?

Baptista thought bird songs shed light on elements of human speech. For instance, young birds learned better from a live tutor than from a tape recording, as do humans. Human words and some bird songs are constructed the same way, linking sounds with individual meanings into a larger piece with its own significance. Of his work, he wrote:

> The many parallels between the structure, geographical variation, and acquisition of human language, on the one hand, and bird song, on the other, make the songs ideal models in studying the evolution of human language . . . I feel confident that exciting discoveries are still to come.[28]

Adult mountain white-crowned sparrow.

Unfortunately, Baptista died in 2000 at only 58, with many investigations left incomplete.

Recently, another ornithologist went back to San Francisco, made recordings of the same white-crowned sparrow populations and compared them to Baptista's original tapes. He found that in the Presidio, no one was singing the Presidio song; they all sang the San Francisco dialect, except a few who sang a hybrid of the two. The Lake Merced dialect was also almost extinct, replaced by the San Francisco song. Looking at the spectrograph, the ornithologist noticed that the San Francisco dialect employed a higher frequency than the other two, so that it could be heard more easily over urban noise.[29] Like their namesake, the Old World sparrows, the white-crowned birds were adapting: accommodating the rush of cars, the growl of airplanes, the overall electric hum. Like people, they were finding the right language for a given geography, a specific neighbourhood. They were adapting to city life.

6 Inheriting the Earth

A jar at the Florida Museum of Natural History contains the corpse of a tiny, dark bird. It looks as though it had dipped its head in ink, which ran down its chest, leaving only the underbelly white. The eye sockets are blank, not filled in with glass as they would be in a taxidermy bird. A silver band circles one leg. The body itself seems shrunken, the pale beak and feet too large for the rest. Though this sparrow was very old when it died, something about the way the head and chest feathers are rumpled and out of place, and its vulnerability alone there in the jar make it seem quite young. Its heart was taken to the University of Georgia. Some strands of its DNA lived on briefly in hybrid chicks. The call, a metallic *tu–weeeeeee*, can be heard on the Internet. But this jar holds the most tangible remains of Orange, the last of the dusky seaside sparrows, which went extinct in 1987.

The dusky seaside sparrow (*Ammodramus maritimus nigrescens*) lived in the salt marshes of south Florida, nesting in salt grass and sand cordgrass, only a foot or so above the ground. The dusky lived on grasshoppers and spiders picked out of the rushes. In the 1960s, NASA built the Kennedy Space Center on Merritt Island, the centre of the dusky seaside sparrow's habitat. With marsh comes mosquitoes, and the stinging insects plagued astronauts and engineers. DDT, sprayed to wipe out mosquitoes, also wiped out many of the little birds, weakening the shells of their eggs. Flooding aimed at ruining mosquito breeding grounds ruined sparrow breeding grounds as well. A state highway paved over a swath of habitat along St John's River. Fires set by ranchers to improve cattle grazing winnowed the dusky's numbers further.[1]

Never abundant and unable to make peace with development like white-crowned and house sparrows, the dusky went quickly. When the number of sparrows was down to thirteen, the Associated Press warned that the dusky might be the first species to go extinct with the Endangered Species Act in place.[2] The act, passed in 1973, required the federal government to protect species and subspecies of plants and animals listed as 'threatened' or 'endangered'. It was considered a conservation milestone, offering hope for many species in trouble, including the dusky, which was listed as 'endangered'. But the sparrows continued to disappear, aided by a lacklustre governmental response to the crisis. Biologist Will Post attributed the carelessness of the Fish

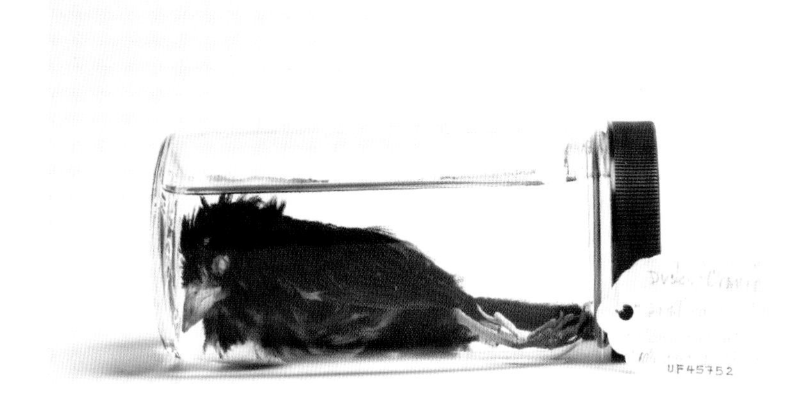

Orange Band, the last dusky seaside sparrow, 2009.

and Wildlife Service to the fact that the bird was obscure and dowdy, not attention-grabbing like a bald eagle or an alligator. The more glamorous creatures were recovering. The dusky was not. 'The small nondescript animals – we can't just afford to go around wiping them out', Post told an Associated Press wire service reporter. 'But people just don't care.'[3]

In 1980, biologists trapped five of the six last known duskies, hoping captivity would keep them safe until some way could be found to stave off extinction. They dubbed them Orange, White, Red, Yellow and Blue, after the colour of their leg bands. All were males, a situation stark in its futility.

First the five were brought to the Florida Fish and Game Commission offices while the hunt continued for a female in the wild. Search after search turned up nothing. From there the birds moved to the Santa Fe Teaching Zoo, and finally to Discovery Island, an attraction at Florida's Walt Disney World. Here they kept company with trumpeter swans and Galápagos tortoises in a display of the rare and imperilled.

In an effort to delay the duskies' complete disappearance, biologists experimented with creating hybrids, breeding the remaining males to Scott's seaside sparrow females. They then hoped to mate some of those offspring with the males, 'back-crossing' until they had a bird that was almost 90 per cent dusky seaside sparrow. Nobody can say the little birds didn't make an effort. Matings with Scott's seaside sparrows produced a number of 50, 75 and 87.5 per cent dusky offspring.[4] Disney World wanted a happy ending.

But slowly, the dusky community was whittled away. Red died of a tumour before reaching Discovery Island. Blue's kidneys failed. Yellow fathered a series of eggs that didn't hatch before he died. By early 1986, White was gone, too. All the hybrids died, either by bashing into the cage or being eaten by rats.

Dusky seaside sparrow (*Ammodramus maritimus nigrescens*).

The remaining bird, Orange Band (who was old, infertile and half blind) became as famous in his own way as Martha, the last of the passenger pigeons, who died in the Cincinnati Zoo in 1914. He was touted as the world's rarest bird, and articles in the *San Francisco Chronicle*, the *Boston Globe* and the *Atlanta Journal and Constitution* featured personality profiles of him. The disappearance of passenger pigeon flocks that only a few decades before had blotted out the sun had been a splash of cold water in the face of the country, awakening it into an age of human-caused extinctions. In contrast, the death of Orange Band occurred after a century of extinctions, and after the Endangered Species Act had been written to prevent just this from happening. Biologists and managers had spent over $2 million buying habitat for duskies.[5] (At the time of the purchase, about 2,000 duskies were left, making the going rate about $1,000 per bird.) They had created the St Johns National Wildlife Refuge to protect them. The birds' disappearance happened in full view of the press and the public and families on vacation at Disney World, who could cluster in front of the dusky seaside sparrow exhibit, even if Orange Band's cage was tucked out of sight. And still, despite all this effort and attention, on the night of 15 June, Orange Band died, taking his race with him.

And when the sparrow fell, everyone noticed. The event was covered by the *New York Times*, the *San Jose Mercury News*, the *Philadelphia Inquirer*. Letters to the editor, columns and headlines referenced the Bible, *Hamlet*, and the gospel hymn 'His Eye is on the Sparrow'. The headline in the *St Petersburg Times* read 'A Sparrow's Fall, The dusky is gone – so what?'[6] 'Sparrow's fall marks "sad" ending',[7] stated *USA Today*. The *Sun-Sentinel* offered an editorial with the headline: 'Fall of Sparrow Symbol of Man's Insensitivity to "Earthmates"'.[8]

The story had a final twist. After Orange Band died, geneticist John Avise took the bird's heart and liver to the University of Georgia for genetic testing. Though it was originally thought to be a full species, the dusky was determined to be only a subspecies in 1973. But Avise wondered if it was even that. After looking at mitochondrial DNA from almost all the subspecies of seaside sparrow, he concluded that the dusky was no different genetically from other subspecies that lived on the east side of Florida. In his paper in *Science*, Avise called Orange 'a routine example of the Atlantic coast phylad of seaside sparrow',[9] and pointed out that if these tests had been done earlier, the dusky would never have qualified for ESA protection.

So what exactly was lost? Was the dusky an emblem of our failure to save an endangered species when all opportunities were available, or a waste of time and money because it wasn't worth saving? What kind of story is this, a tragedy or farce?

It depends on the source. In biology textbooks and academic articles, the dusky's Endangered Species Act listing is often dismissed as a costly error. But in popular culture, the dusky is still shorthand for extinction, a homegrown dodo. In *National Geographic*, columnist Verlyn Klinkenborg uses the example of the sparrow to launch a plea for saving the Act.[10] A few years ago, a Pittsburgh radio station created a series of 'Sounds You Never Heard'. Drivers slowing to exit the Grays Bridge could tune in to hear the dusky seaside sparrows' whirring song and contemplate rarity for a moment before accelerating into their day.[11]

Other New World sparrow species remain on lists of endangered and threatened species in the USA and elsewhere. The Cape Sable seaside sparrow (*Ammodramus maritimus mirabilis*) is a close relative of the dusky. Lighter than other seaside sparrows, the Cape Sable builds a low nest of grasses with sturdy sawgrass

John James Audubon, 'Grasshopper sparrow' (*Ammodramus savannarum*), from *Birds of America* (1827–30).

as its base and softer plants as lining. When the water rises too high, the birds stop breeding. The roughly 3,000 remaining birds mostly live in prairie on either side of Shark River Slough in southern Florida. Water management schemes and agriculture intruding on habitat are shrinking these numbers. Another rare bird, the Florida grasshopper sparrow (*Ammodramus savannarum floridanus*), is named for its insect-like, buzzing song. This bird lives in dry prairie in central Florida and relies on frequent fires that keep the grasslands free of trees and open up bare patches where it can hop around. It digs a hole in the sand for its grass-covered nest near a small saw palmetto or live oak. There are scarcely more than 300.[12]

San Clemente sage sparrow (*Amphispiza belli clementeae*), San Clemente Island, California.

On the other side of the continent, the threatened San Clemente sage sparrow (*Amphispiza belli clementeae*) lives on an island that the Navy uses as a bombing range. Like the dusky and the Florida grasshopper sparrow, the San Clemente sage sparrow is a subspecies, a designation which may not hold up under genetic testing. Joseph Grinnell, an early visitor (the same Grinnell who discovered the Death Valley house sparrows), found the birds thriving among the thorns and cactus spines, and gathered specimens.[13] The bird has a black head and a white circle around the eye, giving it an alert look as it plucks insects from California boxthorn. It nests in boxthorn, too, often surrounded by prickly pear cactus to deter predators.

San Clemente Island is an odd place. A fake Middle Eastern village allows soldiers to train in a realistic-looking market place and school. Cruise missiles blow up island targets. There is no fresh water, except from pools of rain. The fog-bound island,

120 km (75 miles) off the coast of California, is populated by unique species, including the San Clemente loggerhead shrike and the San Clemente island fox. The island night lizard, which also makes its home in prickly pear cactus, lives on several other of the Channel Islands as well, but nowhere else in the world.

Non-native species have also moved in. House sparrows hop along the streets in town and search for seeds by the airport. Nineteenth-century ranchers let their cattle, pigs and goats roam the hills. Ships brought rats and settlers brought cats. These exotic species ate and trampled native plants and some, like the cats, ate sage sparrows. In 1976, scientists estimated there were fewer than 100 San Clemente sage sparrows left.[14] In the 1980s and early 1990s, the government killed many of the goats, and began replanting the island oak, California sage brush and coastal bladderpod. With nesting places growing back, sage sparrows reached a high of about 1,500 in 2002.[15]

But as always with a population this small, persistence depends on nothing going wrong. Killing the goats was easier than shielding chicks from predators. Fires erupting from military operations are a hazard. Avian pox may be infiltrating the population. Recently sage sparrow numbers have started to nosedive again. In 2002, southern California experienced its driest year of the previous 150. Another drought followed in 2006–7. A survey in 2007 found only twelve nests with twenty-one young sparrows.[16] All the island's species respond to rain, but none as much as the sage sparrow, which stops breeding without enough water.

Teegan Doughtery, a biologist with the Institute for Wildlife Studies, placed radio transmitters on recently hatched chicks to determine why so few young survive. The lifespan of each transmitter was three weeks, giving researchers a brief glimpse into the birds' early lives. The reports were sobering. Predators were eating many of the chicks. For the first time, Doughtery documented

A Cuban stamp shows the 'Zapata' Sparrow, *Torreornis inexpectata*, c. 1970.

an endangered San Clemente loggerhead shrike eating a young threatened San Clemente sage sparrow. Her calculations predicted a juvenile survival rate of only 8 per cent, not enough to keep sage sparrows on San Clemente for long.[17]

Internationally, other New World sparrow species are at risk too, particularly those that live on land that has been long ignored but is now coveted by developers or ranchers, who want to burn it for pasture. The Sierra Madre sparrow (*Xenospiza baileyi*) nests in clumps of grass near Mexico City. Worthen's sparrow (*Spizella wortheni*) perches in prickly habitat in high, dry deserts. The Cuban sparrow (*Torreornis inexpectata*) has a big black beak that lets it feast on snails and lizards near the coast. All are on the International Union for Conservation of Nature (IUCN) Red List, labelled as 'endangered'.[18]

If Orange was briefly the rarest bird, alone in his Discovery Island cage, the house sparrow is by far the most common. But even the house sparrow, the seemingly omnipotent house sparrow, is in trouble. Worldwide, the flocks are disappearing. And no one really knows why.

Not long after house sparrows reached heights of population that seemed unbearable to those who found their chirping irritating and wanted faeces-free sidewalks, some observers noted that neighbourhoods were changing. In Denver, Colorado, for example, W. H. Bergtold, a surgeon, found the courthouse square across from his office suspiciously clean and quiet in 1919. He remembered that 'fifteen years ago both its trees and lawn were simply alive with English Sparrows'.[19] Now he might see a handful or a pair, and the nesting boxes he set up for house finches remained undisturbed, whereas before he had had to fight off the sparrows. Bergtold, an amateur naturalist who occasionally wrote articles about robins drunk on honeysuckle berries and the importance of collecting parasites from bird skins, decided to investigate.

He wondered if sparrows might be suffering from a lack of food. '[T]his species lives, especially in the downtown districts, almost exclusively on horse manure',[20] he wrote in an article for *Auk*. At times officials worried that cities would drown in droppings, but the problem no longer loomed so large. Horses were rapidly being replaced by automobiles. Bergtold methodically counted all the horses he saw. He contacted the water company and noted that the number of horses provided with water declined from 1907 to 1917. Records from the Department of Civic Activities showed that street sweepings (mainly manure) per block had decreased almost by half between 1911 and 1919. To Bergtold, the sparrow decline was all to the good. He liked the robins and house finches and was surprised and pleased 'that an advance of civilization can bring about a beneficial change in the biology of a large city'.[21]

House sparrow numbers dipped all over as the horses and all their lightly processed oats left the city, then populations stabilized. But in the 1970s in the United Kingdom, birders noticed that the house sparrows that once had clogged a given park or reliably cluttered up an alley were simply gone. Most of the evidence was circumstantial because the birds had been so common that no one had really bothered to count them. But in Kensington Gardens in London, a 21-year-old man tallied 2,603 house sparrows in 1925.[22] This was already a decline, as at the turn of the century numbers were probably even higher. But the real surprise was that when he counted again, in 2000, there were only eight.[23] Something besides the horses was obviously to blame.

And the birds didn't only vanish in London. In the UK, the bird declined 60 per cent between 1970 and 2000.[24] In northern

A man feeds sparrows in St James's Park, London, 1975.

Logo for World House Sparrow Day.

Italy, a census of specific buildings showed a 50 per cent decline in sparrows over a decade.[25] Sparrows are disappearing in Brussels, Prague, Rotterdam and Moscow. Australia is watching them go. In India, a young man named Mohammed Dilawar is tracking the house sparrow decline, feeding a local population in Nashik, building wooden houses for them, worrying that no one is paying attention. 'It's only ignored because it's common; it has little glamour as compared to other species. There is little awareness with regard to the ecological role it plays', he told an interviewer for *The Hindu*.[26] *Time* magazine selected him as one of its 'Heroes of the Environment' for 2008.[27]

Close relatives of the sparrow, other birds in the *Passer* genus, are also struggling. The Eurasian tree sparrow, a house sparrow cousin which fills its niche of 'the little bird that lives near people' in parts of Asia, is faltering. The tree sparrow is on the Chinese government's protected species list, which outlaws their capture and killing. Anecdotal evidence puts the decline in some areas of China as high as 90 per cent, though unlike in Europe, sparrows are vanishing quickest from rural areas.[28] Still, without a recent anti-sparrow campaign the reason is unclear. Pesticides spread too liberally on farms? Changes in agricultural practices? A fashion for fried sparrows as a snack?

In other countries, scientists are equally stumped. Suggestions include radiation from cell phone towers, radiation from Chernobyl, boys who shoot them with slingshots, high-tech harvesting machines that don't spill much grain, a chemical by-product of unleaded gas, sparrow-proof architecture and the stress of industrial life. It's hard to find anything suitably drastic. Surely nineteenth-century London, when coal smoke saturated the fog, was more polluted than today, yet sparrows were more numerous than ever. Hadn't that been the house sparrow's special genius, to live among bricks and grime and soot? (Some have

House sparrow, from Charles Stanham, *Birds of the British Isles* (1907).

suggested the city is now too clean, robbing the trash-loving birds of their favourite meal.)

In the UK in particular, the public has rallied to protect the city birds that were London's main ornithological event. The birds appear on the Red List of 'Birds of Conservation Concern', and the British Trust for Ornithology raised almost £1 million for their preservation.[29] City parks are cultivating grasses and wildflowers in managed meadows to provide the sparrows with more food. Citizens count and report the birds in their gardens, provide seed and design tempting nest boxes.

Even the Tate Modern has weighed in. In 2004, the art museum hosted an exhibit by Michael Elmgreen and Ingar Dragset of a sparrow trapped between two panes of glass in the window of an otherwise empty gallery. The animatronic bird's death throes – twitching legs, frantically beating heart – were controlled by a computer programme. The exhibit challenged viewers to confront the sparrow's fall on many levels. When asked, the artists and the curator referenced the bird as a representative of a declining species but also set it up as a more abstract symbol. Dragset commented on the cliché of the 'cockney sparrow', and said, 'We wanted to show that working-class culture and working-class pride are dying out, even though there are still as many poor people.'[30] But for many passers-by it was just a dying animal, and many paused at the window, thinking the bird was real and in distress.

In 2000, the British newspaper *The Independent* launched a 'Save the Sparrow' campaign and offered a £5,000 reward to anyone who could solve the mystery.[31] Denis Summers-Smith served as one of the judges.

The prize remained unclaimed for eight years. Finally Kate Vincent, a graduate student at De Montfort University in Leicester, offered the results of her PhD dissertation.[32] She had attached hundreds of bird houses to buildings in and around Leicester, and checked in on the house sparrows that nested there. She found dead chicks as well as eggs that had never hatched. She put leg bands on the young that did hatch, so she could follow their progress. After a while one thing became clear: the young were starving.

As frustrated Americans noted soon after house sparrows arrived, on most days the birds prefer pecking new peas to eating caterpillars, but they do feed their nestlings insects for the first few days. Something about that burst of buggy protein is vital to the nestlings' development. Summers-Smith speculated that lack of insects was why the young at the bottom of the coal mine died, though their parents did fine on scraps from miners' lunches.[33] In her study, Vincent noticed that nestlings fed aphids and spiders did well; those fed on ants and plants died in the nest or fledged with very little fat on their bones, perhaps not enough to survive. Vincent counted aphids and mapped the landscape around nest boxes, noting shrubs, flowerbeds, weed patches and sidewalks. She pulled spider fangs and ant mandibles from the chicks' excrement to see how much of their diet was made up of plants. She also charted levels of pollution from car exhausts.

Only about half the eggs laid produced chicks that eventually flew away, a number well below what would be needed to maintain populations. In addition, chicks raised in spots polluted with nitrogen oxide, a byproduct of burning gasoline or coal, tended to be scrawny, and scrawny chicks often didn't survive. Over the course of the three-year study, house sparrow populations in Leicester declined by almost 30 per cent.[34]

As she monitored the sparrows, Vincent found that people were eager to help, putting nest boxes on their houses, offering their opinions. 'Everyone's got something to say about house sparrows, even if they're not a scientist', she says. As a conservation issue, it hits home more than a bird at risk in Borneo. 'It's a way to engage people . . . It's more tangible than some species they have never seen or are never likely to see.'[35]

Vincent's study swaps one question for another. Where are the insects? Some pollutant associated with city life might be killing them off. The increased paving over of city parks and weed lots and front lawns could be taking a toll. Even without a definite answer, *The Independent* is now educating its readers about sparrow conservation and the need for insects in cities. The BBC runs stories like 'Making a Garden Sparrow-friendly', urging native plants and a slightly messy aesthetic, rather than tidy lawns and hedges.[36] On her website Vincent includes tips on how to build house sparrow nesting boxes. All this is a long way from the nineteenth century, when cities imported house sparrows to eat the insects; now cities are promoting insects so they can have more house sparrows.

Others are investigating the sparrow mystery, too, mulling over alternate solutions. Ornithologist Christopher Bell was working at London Zoo – one of the few places in the city with

Attributed to Ren Renfa (Jen Jen-fa), detail from *Hawk Killing Sparrow*, c. 1300, ink and colour on silk.

a decent sparrow population – when he noticed that the remaining birds fed inside the animal cages. Bell saw sparrowhawks lurking nearby, picking off the birds when they ventured outside the bars. Like many other areas, the Zoo started a house sparrow conservation programme, erecting nestboxes and setting up feeding stations. The sparrowhawks became even more bold, going so far as to chase the sparrows into the women's restroom and devouring them there. A sudden explosion in the bushes indicated a hawk attack. 'We used to find sparrows in bits', he says.[37]

A diminutive predator with a slate-grey or brown back, a rust-coloured chest and startling yellow eyes, the sparrowhawk, *Accipiter nisus*, was a relatively new presence at the zoo. It had long been a favourite of falconers, noted in particular for its skill in chasing down small birds, flying fast and low, crashing into a hedgerow to secure its prey. But the species had been decimated in the 1950s by the use of chlorinated hydrocarbon pesticides, like aldrin and DDT, which weakened sparrowhawk eggs. Even before that, gamekeepers protecting their pheasants kept sparrowhawk numbers low by shooting them. But after the pesticide use was reduced and sparrowhawks were protected from hunting in the early 1960s, sparrowhawks in Britain staged a comeback, growing from less than 10,000 pairs in the 1960s to more than 32,000 pairs thirty years later.[38] Before long, city dwellers could watch them soaring over tall buildings, pigeons fleeing as they came.

Intrigued at a possible correlation between more sparrowhawks and fewer sparrows, Bell and colleagues from Cambridge University and the British Trust for Ornithology analysed data from the British Trust for Ornithology's weekly counts of birds at feeding stations each winter. They found that sparrows began to disappear at about the same time as sparrowhawks began showing up at feeders year after year. The effect was most pronounced in urban areas where, according to Bell, sparrows had adjusted to life without predators, growing increasingly confident and bold. 'They got clobbered because they didn't know what hit them', Bell says.

A few large European cities still have dense, dauntless clumps of sparrows, but possibly not for long. 'If you go to Paris and you go to Notre Dame, you can still get that authentic urban sparrow experience of sparrows feeding on your hand', Bell says. In the past few years, though, sparrowhawks have moved in and started to breed. He adds, 'In five or ten years' time, you won't see any sparrows there. That's my prediction.'[39]

These two stories – Vincent's declining insects and Bell's increasing sparrowhawks – offer markedly different visions of not just the sparrow problem, but the state of the environment. In one, cities are afflicted by a perplexing disease, not yet fully diagnosed, that cripples insects, one of the hardiest life forms. In the nineteenth century, house sparrows were painted gold and

The bird charmer at the Tuileries, Paris, from *Le Petit Journal*, 22 January 1911.

sold as canaries to buyers who fantasized that they could sing. But now the sparrow is referenced as another kind of canary: the canary in the coal mine, brought underground to swoon in toxic air, giving miners the chance to escape.[40] Scientists wonder if the house sparrows and the insects that feed them are failing under some unknown aspect of the urban environment that will soon affect people, too. Perhaps the vanishing sparrows are telling us something about the state of the cities that we can't yet perceive with our own senses.

In the competing story, nature is healing itself. The ability to support predators is one of the tests of a restored ecosystem. Laws to protect wildlife have started to work; cities are clean enough that they can host not just traditional 'trash birds' but a

A boy near Notre Dame, Paris, feeding urban house sparrows.

Le Petit Journal

ADMINISTRATION
61, RUE LAFAYETTE, 61
Les manuscrits ne sont pas rendus
On s'abonne sans frais
dans tous les bureaux de poste

5 CENT. SUPPLÉMENT ILLUSTRÉ 5 CENT.

22me Année — Numéro 1.053

DIMANCHE 22 JANVIER 1911

ABONNEMENTS

	SIX MOIS	UN AN
SEINE et SEINE-ET-OISE..	2 fr.	3 fr. 50
DÉPARTEMENTS..........	2 fr.	4 fr. »
ÉTRANGER................	2 50	5 fr. »

UN TYPE PARISIEN
LE CHARMEUR D'OISEAUX DES TUILERIES

bevy of critters that may demand more from their environments. It's a dangerous tale, because it may lead to a backlash against birds of prey just as they are finding their footing again, and anti-predator campaigns always find an eager following. Ultimately, though, it's an optimistic one. Sparrow populations, artificially high as the result of the persecution of their predators, will find a natural level. Bell says this may already be happening in rural areas in the West where sparrowhawk numbers surged early and the sparrow decline has slowed. 'You get an equilibrium between prey and predator', says Bell. 'I think sparrows will come back.'[41]

An oversized sculpture of a sparrow, Vancouver, British Columbia, Canada. The sculpture at the former athletes' village for the Vancouver Olympics is one of a pair titled *The Birds* by local artist Myfanwy MacLeod.

Reading about all this anguish over the house sparrow, one can't help but wonder: 'What would Elliott Coues say?'

He might point out, as he did during the Sparrow War, that whether you celebrate or revile a creature depends its relationship to its environment. In the U.S., where house sparrows are not native, the birds are for the most part still considered pests along the lines of rats and cockroaches. While almost all wild birds are protected from random killing by U.S. law, you can do whatever you like to exotic species like starlings, pigeons or house sparrows. Evidence shows house sparrow declines in the U.S. as well, though their disappearance is often deemed cause for celebration.[42]

Skirmishes rooted in the Sparrow War still break out on occasion. One hundred and ten years after the meeting in Boston, a shopping centre in West Windsor, New Jersey, planned to spread chemicals to rid itself of the sparrows that coated the sidewalk in front of the grocery store and pharmacy with droppings. Concerned residents called the health department. They contacted People for the Ethical Treatment of Animals, an animal rights organization. Teenage members of 'Save Our Sparrows' collected signatures on a petition outside a pizza shop and the management agreed to use netting to discourage the sparrows instead.[43] A sculpture exhibit at the University of Connecticut featured a very different take on a dead sparrow from that offered by the Tate Modern. A sculpture by Randall Nelson titled 'The Birdwatcher's Verdict' showed a sparrow in a noose by the inscription 'The bird got what it deserved.'[44] Students protested and asked that it be moved to a less prominent spot on campus.

At www.sialis.org, a website for bluebird lovers, house sparrows are listed as 'the number one enemy of bluebirds' because sparrows will drive bluebirds from their nests, killing the parents and young. The site shows graphic images of bluebirds

Sparrow trap.

decapitated by house sparrows. Visitors can learn how to install a shiny 'sparrow spooker' over their bird houses or put a 'magic halo' of wires by their feeder to scare off house sparrows. They are advised to put a plastic snake near a house sparrow nest to chase the parents away or to cut away bushes or ivy where the birds might rest. Destroying nests, breaking eggs or dipping them in oil so they won't hatch are other suggestions. Since none of these methods are fail safe (sparrows can rebuild nests in no time and have been known to nest right on top of the toy snake), readers are told how to trap them, gas them with CO_2, break their necks and shoot them. They can then freeze the bodies and donate them to raptor rehabilitation centres.

A 1930s leaflet on English sparrow control.

This animosity against exotics means that Americans are taking more pride in their native sparrows. The poet Elizabeth

ENGLISH
SPARROW
CONTROL
LEAFLET
No. 61

John James Audubon, 'Henslow's sparrow' (*Ammodramus henslowii*), from *Birds of America* (1827–30).

(*opposite*) Savannah sparrow on the Kenai Peninsula, Alaska.

Bishop, like Emily Dickinson, early on used the sparrow in its classical European sense in 'Three Valentines':

> Love with his gilded bow and crystal arrows
> Has slain us all,
> Has pierced the English sparrows
> Who languish for each other in the dust.[45]

These are Old World literary sparrows, associated with Venus and lust. Bishop later adopts New World sparrows as her trope, highlighting their music, comparing to a poet's. Mourning the death of a friend in 'North Haven', she writes 'the White-throated Sparrow's five-note song,/ pleading and pleading, brings tears to the eyes.'[46]

Because house sparrows' lives are interwoven with our own in immediate, everyday ways, it's easy to see how our fates might be bound together and feel that their loss might be our loss. We share the habitat of the city. The impact of their disappearance is felt on a trip to the park or a walk downtown. Their high profile brings media attention and helps raises money for their assistance.

On the other hand, many endangered New World sparrows live most of their lives in obscurity on low-cost real estate and their decline is only noticed when the remnants of their whole species could fit in a train station locker. They haven't been kept as pets, or had Roman poems written about them, or been praised and damned in newspaper editorials. Their perceived value is not even half a farthing. It takes a leap of imagination to engage with their concerns. But when the little birds of the salt marsh, or the mixed thicket of boxthorn and prickly pear, or the bunchgrass meadows on Mexican mountain slopes are gone, it means those ecosystems are gone, too, with all their plants, reptiles and insects. And the world is impoverished.

Several months ago, I was chatting at a Christmas party with the mother of an old friend. I told her about the sparrow, about the Bible, about God's care for even the cheapest forms of life, about the way people, when feeling insignificant, have taken the idea that 'His Eye is on the Sparrow' to heart. They've understood the sparrow story in the Gospel of Matthew as a promise about the importance of the unique self. 'Well,' she said. 'I hope it's about conservation.' Humble species need our attention, she added, not just overlooked individuals, human or avian. It should be a reminder to us to keep our eye on the sparrows, all of them. And even if that's not what Matthew originally meant so long ago, maybe it's the interpretation we need today.

Timeline of the Sparrow

1 million BCE

A sparrow ancestor begins to radiate out from the African tropics

400,000 BCE

A species recognizable as *Passer domesticus* lives near early humans in the Middle East

1,000 BCE

House sparrows flutter around cooking pots in what is now Sweden

64 BCE

Catullus mourns his lover's sparrow

1743

Tommy Thumb's Pretty Song Book asks 'Who Killed Cock Robin?' (The sparrow did)

Early 19th century

Sparrow Clubs in UK pay bounties on sparrows

1827–38

Audubon publishes *Birds of America*

1853

First successful release of house sparrows into the United States

1917

Joseph Grinnell discovers house sparrows living in Death Valley, California

1958

Chairman Mao orders the Chinese to exterminate sparrows

Early 1970s

Biologists begin studying white-crowned sparrows' song

1970s

House sparrow numbers in Europe decline

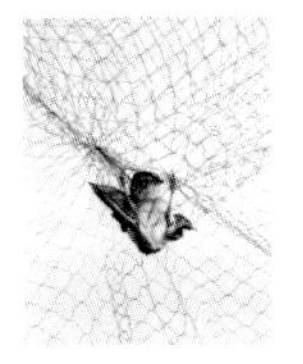

1505–10

Jane mourns her pet in John Skelton's poem 'Phyllyp Sparowe'

1525/6

First Bible in English states that two sparrows were sold for a farthing

1602

Hamlet finds 'special providence in the fall of a sparrow'

1650–1700

Settlers in the the New World dub unfamiliar small brown birds 'sparrows'

1863

House sparrows introduced to Australia

1898

Hermon Carey Bumpus uses sparrows to study natural selection

1900

In the USA, the Lacey Act bans importation of house sparrows

WAR OPENS.

Raid Begun Upon the Sparrows.

Nests Torn Down on Common.

Mayor's Response to a Protest.

1904

Civilla Martin writes 'His Eye is on the Sparrow'

1973

Endangered Species Act passes in the US

1987

The dusky seaside sparrow goes extinct

2000

The *Independent* offers a £5,000 reward for an answer to the mystery of why house sparrows are disappearing

2002

House sparrow joins list of UK's 'Birds of Conservation Concern'

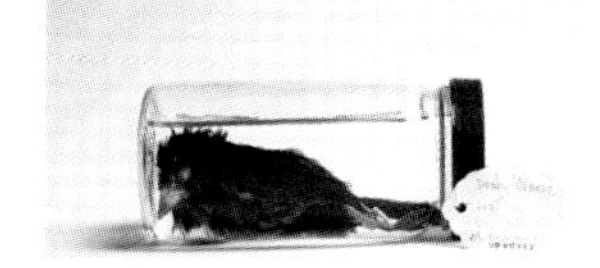

References

INTRODUCTION

1 William Shakespeare, *Hamlet*, V.ii.219–20.
2 Spencer Trotter, 'An Inquiry into the Current English History into the Names of North American Land Birds', *Auk*, XXVI/4 (1909), pp. 346–63.
3 Sei Shonagon, *The Pillow Book*, trans. Arthur Waley (Whitefish, MT, 2005), p. 123.
4 Elliott Coues, 'Avifauna Columbiana', in *The House Sparrow*, ed. J. H. Gurney (London, 1885), pp. 60 and 61.
5 Elliott Coues, 'The Ineligibility of the European House Sparrow in America', *American Naturalist*, XII/8 (1878), p. 503.
6 Samuel Christian Schmucker, *The Meaning of Evolution* (New York, 1922), p. 84.
7 Denis Summers-Smith, *The House Sparrow* (London, 1963), p. 5.
8 'Sparrow', *The Oxford English Dictionary* (Oxford, 1961).
9 'Backyard Birds of Winter in Nova Scotia', http://museum.gov.ns.ca/mnh/nature/winbirds/colour/c14.htm.
10 Norman Boucher, 'Whose Eye is on the Sparrow?' *New York Times Magazine*, 13 April 1980, p. SM13.
11 Emily Dickinson, 'Victory comes late', ll. 13–14.
12 Emily Dickinson, 'Her breast is fit for pearls', ll. 5–8.

1 LITTLE BROWN JOBS

1 'Animal Life in Mines', *The State*, 1 March 1915.
2 Denis Summers-Smith, 'House Sparrows Down Coal Mines', *British Birds*, LXXIII (1980), pp. 325–7.
3 Ted Anderson, *Biology of the Ubiquitous House Sparrow* (Oxford, 2006), p. 153.
4 J. W. Parker, 'Additional Records of House Sparrows Nesting on Raptor Nests', *Southwestern Naturalist*, XXVIII (1982), pp. 240–41.
5 Thomas Gentry, *The House Sparrow at Home and Abroad* (Philadelphia, PA, 1878), p. 43.
6 Aristotle, 'On Longevity and Shortness of Life', *Aristotle's Psychology*, trans. William Alexander Hammond (New York, 1902), parts 5–6, p. 265.
7 Pliny the Elder, *Natural History*, Book X, section 36.
8 J. H. Wetton et al., 'Demographic Study of a Wild House Sparrow Population by DNA Fingerprinting', *Nature*, CCCXXVII/6118 (1987), pp. 147–9.
9 J. Veiga, 'Infanticide by Male and Female House Sparrows', *Animal Behaviour*, XXXIX (1990), pp. 496–502.
10 Denis Summers-Smith, *The House Sparrow* (London, 1963), p. 28.
11 A. P. Gavett and J. S. Wakeley, 'Diets of House Sparrows in Urban and Rural Habitats', *Wilson Bulletin*, XCVIII (1986), pp. 137–44.
12 R. Breitwisch and M. Breitwisch, 'House Sparrows Open an Automatic Door', *Wilson Bulletin*, CIII (1991), pp. 725–6.
13 'Usefulness of Birds: An Official Investigation of the Feathered Tribes of the United States', *The Troy Weekly Times*, 26 November 1885, p. 1.
14 H. J. Brockman, 'House Sparrows Kleptoparasitize Digger Wasps', *Wilson Bulletin*, XCII (1980), pp. 394–8.
15 The number of 27 species comes from an interview of J. Denis Summers-Smith by the author (November 2009). In *The Sparrows* (Calton, Staffordshire, 1988), his study of the genus *Passer*, Summers-Smith lists twenty species.

16 H. Stevens, 'Notes on the Birds of the Sikkim Himalayas', Part v, *Journal of the Bombay Natural History Society*, xxx (1925), pp. 352–79.
17 Denis Summers-Smith, *The Sparrows* (Staffordshire, 1988), pp. 221–2.
18 Ibid., pp. 276–96.
19 Ted Anderson, *Biology of the Ubiquitous House Sparrow* (Oxford, 2006), p. 9.
20 P.G.P. Ericson et al., 'The Earliest Record of House Sparrows (*Passer domesticus*) in Northern Europe', *Journal of Archeological Science*, xxiv (1997), pp. 183–90.
21 Arturo Morales Muñiz et al., 'Of Mice and Sparrows', *International Journal of Osteoarcheology*, x/2 (1995), pp. 127–38.
22 Lynn B. Martin ii and Lisa Fitzgerald, 'A Taste for Novelty in Invading House Sparrows, *Passer domesticus*', *Behavioral Ecology*, xvi/4 (2005), pp. 702–7.
23 'Emberizidae: Emeberizid Finches', *Encyclopedia of Life*, online at www.eol.org/pages/7553, accessed 4 February 2011.
24 John Josselyn quoted in Elsa Guerdrum Allen, 'The History of American Ornithology before Audubon', *Transactions of the American Philosophical Society*, n.s., xli/3 (1951), p. 548.
25 John Josselyn, *New-Englands Rarities Discovered* (London, 1674), p. 12.
26 John Lawson, *A New Voyage to Carolina* (London, 1709), p. 144.
27 Ibid., p. 146.
28 Ibid., p. 135.
29 Mark Catesby, *Natural History of Carolina, Florida and the Bahama Islands* (London, 1731–43), plates 35 and 37.
30 John James Audubon, *Birds of America* (New York, 1840). There are references to finches throughout.
31 Celia Thaxter, 'The Song Sparrow', *Atlantic Monthly*, xxxii (1873), p. 539.

2 SOLD FOR TWO FARTHINGS

1 Louisa May Alcott, *Hospital Sketches* (Boston, MA, 1863), p. 76.
2 Nawal Nasrallah, *Annals of the Caliphs' Kitchens: Ibn Sayyar al-Warraq's Tenth-century Baghdadi Cookbook* (Leiden, 2007), p. 328.
3 Laura Hobgood-Oster, *Holy Dogs and Asses* (Champaign, IL, 2008).
4 William Shakespeare, *Hamlet*, V.ii.219–23.
5 Heinrich Cornelius Agrippa, *Three Books of Occult Philosophy* (London, 1651), p. 114.
6 Homer, *Iliad*, II, ll. 305–19.
7 Aristotle, *History of Animals*, book IX, part 29.
8 John Latham quoted in Elsa Guerdrum Allen, 'The History of American Ornithology before Audubon', *Transactions of the American Philosophical Society*, n.s., XLI/3 (1951), p. 496.
9 William Shakespeare, *King Lear*, I.iv.215–16.
10 William Shakespeare, *1 Henry IV*, V.i.59–66.
11 Anon., *Guy Earl of Warwick*, quoted in *Shakespeare, Marlowe, Jonson*, ed. Takashi Kozuka and J. R. Mulryne (Hampshire, 2006), p. 129.
12 Civilla Martin quoted in Lindsay Terry, *Stories Behind Fifty Southern Gospel Favorites* (Grand Rapids, MI, 2002), p. 172.
13 John Lockwood Kipling, *Beast and Man in India* (London, 1891), pp. 52–3.
14 Patricia Bjaaland Welch, *Chinese Art* (North Clarendon, VT, 2008).
15 Felice Fischer's translation is from correspondence with the author (July 2009).
16 Felice Fischer, 'Japanese Buddhist Art', *Philadelphia Museum of Art Bulletin*, LXXXVII/369 (1991), p. 21.
17 J. F. Handlin Smith, 'Liberating Animals in Ming-Qing China: Buddhist Inspiration and Elite Imagination', *Journal of Asian Studies*, LVIII/1 (1999), pp. 51–84.
18 Sheldon Lou, *Sparrows, Bedbugs, and Body Shadows* (Honolulu, HI, 2005), p. 37.
19 Ibid., p. 41.

20 Han Suyin, 'The Sparrow Shall Fall', *New Yorker*, 10 October 1959, pp. 43–50.
21 Ibid., p. 49.
22 Ibid., p. 50.
23 Ibid.
24 Ibid.
25 Lou, *Sparrows, Bedbugs, and Body Shadows*, pp. 45–6.
26 Ibid., p. 49.
27 S. Karnow, *Mao and China* (New York, 1990), p. 91.
28 Judith Shapiro, *Mao's War Against Nature* (Cambridge, 2001), p. 88. After house sparrows invaded the U.S. in the nineteenth century, many studies purported to show that insects made up a relatively small part of their diet. Perhaps the rise in insects after the eradication was only attributed to, but not caused by, the lack of sparrows or maybe in some circumstances sparrows can have a significant impact on insect populations.
29 'It's Crane v. Sparrow in China's National Bird Search', at http://blogs.wsj.com/chinarealtime/2008/09/10/its-crane-v-sparrow-in-chinas-national-bird-search/tab/article, accessed 4 February 2011.

3 DEAD AND DIRTY BIRDS

1 Catullus, trans. C. H. Sisson, 3, 'Time for mourning, Loves and Cupids', in *The Poetry of Catullus* (New York, 1966), ll. 11–12.
2 H. D. Jocelyn, 'On Some Unnecessarily Indecent Interpretations of Catullus 2 and 3', *American Journal of Philology*, CI/4 (1980), pp. 421–41.
3 Ibid., p. 441.
4 Ibid., p. 424.
5 Sappho, trans. Bliss Carman, 1, 'O Aphrodite', in *Sappho: One Hundred Lyrics* (Boston, MA, 1907), ll. 11–12.
6 Geoffrey Chaucer, 'Parliament of Fowles', l. 351.
7 Geoffrey Chaucer, 'Prologue', *Canterbury Tales*, l. 626.

8 William Shakespeare, *Measure for Measure*, III.ii.175–6.
9 John Donne, 'Epithalamion', l. 7.
10 John Donne, 'The Progresse of the Soule', ll. 208–11.
11 John Skelton, 'Phyllyp Sparowe', l. 348.
12 Ibid., ll. 171–72.
13 Jacob Grimm and Wilhelm Grimm, *The Brothers Grimm: The Complete Fairy Tales* (Ware, Hertfordshire, 1997), p. 281.
14 A. B. Mitford, *Tales of Old Japan* (London, 1888), p. 174.
15 Ibid.
16 Joel Chandler Harris et al., *The Complete Tales of Uncle Remus* (Boston, MA, 2002), p. 57.
17 Hans Christian Andersen, *The Fairy World* (New York, 1877), p. 110.
18 Ibid., p. 113.
19 Ibid., p. 115.
20 Flora Annie Steel, *Tales of the Punjab: Told by the People* (London, 1973), no. 18, pp. 100–7.
21 Charles Swainson, *Provincial Names and Folk Lore of British Birds* (London, 1885), p. 16.
22 W.R.S. Ralston, *Russian Folk-Tales* (London, 1873), p. 331.
23 Iona and Peter Opie, *Oxford Dictionary of Nursery Rhymes* (Oxford, 1951), p. 130.
24 Ibid.
25 Ibid., pp. 130–32.
26 Ibid., p. 131.
27 Ibid., pp. 129–30.
28 William Blake, 'The Blossom', ll. 1–12.
29 Rodney M. Baine and Mary R. Baine, 'Blake's "Blossom"', *Colby Quarterly*, XIV/1 (1978), pp. 22–7.
30 William Blake, 'Auguries of Innocence', ll. 5–6.
31 Ibid., ll. 56–8.

4 THE SPARROW WAR

1 'City Affairs', *The North American*, 27 February 1847, p. 2.
2 Ibid.
3 Walter B. Barrows, *The English Sparrow (Passer domesticus) in North America* (Washington, DC, 1889), p. 17.
4 'A Crusade Against Sparrows', *The Sun*, 23 March 1899, p. 4.
5 Joshua L. Rosenbloom, 'The Extent of the Labor Market in the United States, 1870–1914', *Social Science History*, XXII/3 (1998), p. 289.
6 C.S., 'Street Trees and English Sparrows', *Colman's Rural World*, XXII/11 (1869), p. 166. This is only one of many examples.
7 Elliott Coues, '*Avifauna Columbiana*', in *The House Sparrow*, ed. J. H. Gurney (London, 1885), p. 60.
8 John Oscar Skinner, 'The House Sparrow', *Annual Report, Smithsonian Institution* (1904), p. 424.
9 Fred Mather, 'Old World Nuisance', *Forest and Stream*, XVIII/3 (1881), p. 46.
10 'Another Sparrow Poem', *Forest and Stream*, XVII/7 (1881), p. 126.
11 William Cullen Bryant, 'The Old World Sparrow', *The Poetical Works of William Cullen Bryant* (New York, 1910), p. 373.
12 'The Cincinnati Acclimatization Society', *Forest and Stream*, I/16 (1873), p. 249.
13 Bryant, 'The Old World Sparrow', p. 373.
14 Ibid.
15 Kim Todd, 'Botanically Correct: A New Language is Needed to Win the Day for Native Species', *Grist Magazine*, at www.grist.org/article/correct, accessed 9 February 2011.
16 Georges-Louis Leclerc, Comte de Buffon, quoted in Keith Thomson, *The Legacy of the Mastodon* (New Haven, CT, 2008), p. 38.
17 Colonel William Rhodes, 'Imported Birds for Our Woods and Parks', *Forest and Stream*, VIII/11 (1877), p. 165.
18 Barrows, *The English Sparrow (Passer domesticus) in North America*, pp. 19–20.

19 C.S., 'Street Trees and English Sparrows', p. 166.

20 'The Sparrows' Home', *Harper's Weekly*, 3 April 1969, p. 214.

21 J. L. Long, *Introduced Birds of the World* (Newton Abbot, 1981), pp. 373–82.

22 William Cowper quoted in *Proceedings of the Boston Society for Natural History*, XI (1867), p. 158.

23 L. Wing, 'Spread of the Starling and the English Sparrow', *Auk*, LX (1943), p. 79.

24 H. C. Yarrow cited in Paul Russell Cutright and Michael J. Brodhead, *Elliott Coues: Naturalist and Frontier Historian* (Champaign, IL, 2001), p. 145.

25 Ibid., pp. 463, 473.

26 Elliott Coues, 'English Sparrows', *American Naturalist*, VIII (1874), p. 436.

27 Elliott Coues, 'The Ineligibility of the European House Sparrow in America', *American Naturalist*, XII/8 (1878), p. 499.

28 Elliott Coues, 'Documents in the Bendire Business', *Osprey*, II (1897), pp. 22–3.

29 Coues, 'The Ineligibility of the European House Sparrow in America', p. 505.

30 'Our Speechless Slaves', *Cincinnati Daily Gazette*, 12 December 1873, p. 2.

31 Henry Bergh, 'Letter from Mr Bergh', *Popular Science Monthly*, XV (1879), p. 409.

32 Ibid.

33 Patricia Adair Gowalty, 'House Sparrows Kill Eastern Bluebirds', *Journal of Field Ornithology*, LV/3 (1984), pp. 378–80.

34 Henry Ward Beecher, 'Star Paper', *Christian Union*, XVI/6 (1877), p. 103.

35 Ibid.

36 Wing, 'Spread of the Starling and the English Sparrow', p. 79.

37 Coues, 'The Ineligibility of the European House Sparrow in America', p. 500.

38 'The Butcher-bird', *Duluth Daily Tribune* (18 October 1881), p. 3.

39 'Birds Acquiring New and Brutal Habits', *Daily Constitution*

(10 June 1873), p. 1.

40 'Two Foes of the Sparrow', *Kansas City Star* (4 February 1888), p. 2.

41 'Springfield', *Springfield Daily Republican* (17 January 1877), p. 6.

42 Olive Thorne Miller, 'A Ruffian in Feathers', *Atlantic Monthly*, LX (1885), p. 490.

43 Pele Mele, 'A Chirp about Sparrows', *New Monthly Magazine*, CVL (1869), p. 579.

44 J. Hinton, 'A Meditation on Skeletons', *Cornhill Magazine*, XII (1863), p. 622.

45 Miller, 'A Ruffian in Feathers', p. 490.

46 Ibid., p. 492.

47 'The English Sparrow Nuisance', *Forest and Stream*, XVI/18 (1881), p. 347.

48 James T. Bell, 'Hints on Sparrow Destruction', *Forest and Stream*, XXIX/15 (1887), p. 283.

49 Wilson Flagg, 'How to Kill the English Sparrow', *Forest and Stream*, XX/26 (1883), p. 503.

50 Mather, 'The Old World Nuisance', p. 46.

51 Ibid.

52 'Another Sparrow Poem', p. 126.

53 Ernest Thompson Seton, 'A Street Troubadour', *Lives of the Hunted* (New York, 1901), p. 125.

54 'About Sparrows', *Scientific American*, XXXVII/6 (1877), p. 80.

55 Thomas Gentry, *The House Sparrow at Home and Abroad* (Philadelphia, PA, 1878), p. 34.

56 Ibid., p. 77.

57 Duke of Argyll, 'First Impressions of the New World', *Fraser's Magazine*, XXII (1880), p. 48.

58 Ibid.

59 'Notes and News', *Auk*, XVI/2 (1899), p. 215.

60 'The So-Called Sparrow War in Boston', *Bird Lore*, I (1899), p. 137.

61 'War Opens', *Boston Journal*, 14 March 1899, p. 2.

62 'Sparrow War, the Butcher Bird Suggested as an Ally', *Boston Sunday Journal*, 12 March 1899, p. 13.

63 'War Opens', p. 2.
64 'Mayor Hears Sparrows' Foes Tell their Stories', *Boston Journal*, 18 March 1899, p. 1.
65 Elliott Coues, 'Dr Coues' Column', *Osprey*, I (1897), p. 124.
66 Wing, 'Spread of the Starling and the English Sparrow', p. 79.

5 THE FALL OF A SPARROW IN PROVIDENCE

1 Hermon Carey Bumpus, 'Elimination of the Unfit as Illustrated by the Introduced Sparrow, *Passer domesticus*', *Biol. Lectures: Woods Hole Marine Biological Laboratory* (Boston, MA, 1898), p. 209.
2 Ibid., p. 219.
3 Ibid., p. 216.
4 Ibid., p. 208.
5 'Friendly Purple Martin Competitors', at www.chuckspurplemartinpage.com/compete.htm.
6 Joseph Grinnell, 'The English Sparrow has Arrived in Death Valley', *American Naturalist*, LIII/628 (1919), p. 469.
7 Ibid.
8 Ibid., p. 472.
9 Richard F. Johnston and Robert K. Selander, 'House Sparrows: Rapid Evolution of Races in North America', *Science*, n.s., CXLIV/3618 (1964), pp. 548–50.
10 William Monahan, 'Wing Microevolution in the House Sparrow Predicted by Model of Optimized Wing Loading', *Condor*, CX/1, 2008), pp. 161–6.
11 Lynn B. Martin II and Lisa Fitzgerald, 'A Taste for Novelty in Invading House Sparrows, *Passer domesticus*', *Behavioral Ecology*, XVI/4 (2005), pp. 702–7.
12 András Liker and Veronika Bokony, 'Larger Groups are More Successful in Innovative Problem Solving in House Sparrows', *PNAS*, CVI/19 (2009), pp. 7893–8.
13 Kevin J. McGraw et al., 'Social Environment during Molt and the Expression of Melanin-Based Plumage Pigmentation in Male House Sparrows (*Passer domesticus*)', *Behavioral Ecology and*

Sociobiology, LIII/2 (2003), pp. 116–22.

14 Kate Vincent, 'Investigating the Causes of the Decline of the Urban House Sparrow *Passer domesticus* Population in Britain', PhD thesis, De Montfort University, Leicester (2005), p. 3.

15 Denis Summers-Smith, interview with the author (December 2010).

16 Ibid.

17 Luis Baptista, *Song Dialects and Demes in Sedentary Populations of the White-crowned Sparrow (Zonotrichia leucophrys nuttalli)* (Berkeley, CA, 1975), p. 1.

18 Luis Baptista, 'What the White-crowned Sparrow's Song can Teach Us about Human Language', *Chronicle of Higher Education*, 7 July 2000, p. B8.

19 Baptista, *Song Dialects and Demes in Sedentary Populations of the White-crowned Sparrow*.

20 Luis F. Baptista and Martin L. Morton, 'Song Learning in Montane White-crowned Sparrows: From Whom and When', *Animal Behaviour*, XXXVI/6 (1988), pp. 1753–64.

21 Paul W. Sherman and Martin L. Morton, 'Extra-Pair Fertilizations in Mountain White-Crowned Sparrows', *Behavioral Ecology and Sociobiology*, XXII/6 (1988), p. 413.

22 Andrea Crino, interview with the author (June 2009).

23 Ibid.

24 Ibid.

25 Brett Klaassen van Oorschot, interview with the author (June 2009).

26 Andrea Crino, interview with the author (June 2009).

27 Ronald Wimberley, 'Population Takes a Turn', *News and Observer*, 27 May 2007, p. A15.

28 Baptista, 'What the White-crowned Sparrow's Song can Teach us about Human Language'.

29 David Luther and Luis Baptista, 'Urban Noise and the Cultural Evolution of Bird Songs', *Proceedings of the Royal Society B: Biological Sciences*, CCXXVII/1680 (2010), pp. 469–73.

6 INHERITING THE EARTH

1 Mark Jerome Walters, *A Shadow and a Song* (Post Mills, VT, 1992). Much of this chapter's information about the decline of the dusky seaside sparrow comes from this excellent source.

2 Rick Spratling, 'Sparrows Nesting Near Spaceport Face Extinction', *Associated Press*, 12 January 1980.

3 Ibid.

4 Mark Jerome Walters, *A Shadow and a Song* (Post Mills, VT, 1992).

5 Ibid., p. 122.

6 Jeff Klinkenberg, 'A Sparrow's Fall, The Dusky is Gone – So What?' *St Petersburg Times*, 16 August 1987, p. 1F.

7 Stephan Stern, 'Sparrow's Fall Marks "Sad" Ending', *USA Today*, 18 June 1987, p. 1A.

8 Tom Sander, 'Fall of Sparrow Symbol of Man's Insensitivity to "Earthmates"', *Sun-Sentinel*, 26 June 1987, p.15A.

9 John Avise, 'Molecular Genetic Relationships of the Extinct Dusky Seaside Sparrow', *Science*, n.s., CCXLIII/4891, (1989), p. 648.

10 Verlyn Klinkenborg, 'Last One', *National Geographic*, CCXL/1 (2009), p. 82.

11 'Homestead Radio offers Sound of a Sparrow', www.thepittsburghchannel.com/news/14295064/detail.html, 8 October 2007, accessed 17 February 2011.

12 Bill Pranty and James W. Tucker Jr, 'Ecology and Management of the Florida Grasshopper Sparrow', in *Land of Fire and Water: The Florida Dry Prairie Ecosystem. Proceedings of the Florida Dry Prairie Conference*, ed. Reed F. Noss (De Leon Springs, FL, 2006), p. 191.

13 Joseph Grinnell, *Report on the Birds recorded during a Visit to the Islands of Santa Barbara, San Nicolas and San Clemente in the Spring of 1897* (Pasadena, CA, 1897).

14 Sara A. Kaiser et al., 'Population Monitoring of the San Clemente Sage Sparrow – 2007', Final Report. Unpublished report prepared by the Institute for Wildlife Studies for the United States Navy (San Diego, CA, 2007), p. 1.

15 Ibid., p. 2.

16 Ibid., p. iv.
17 Teegan Doughtery, interview with the author (November 2009).
18 IUCN Red List available online at www.iucnredlist.org, accessed 17 February 2011.
19 W. H. Bergtold, 'The English Sparrow (*Passer domesticus*) and the Motor Vehicle', *Auk*, XXXVIII/2 (1921), p. 244.
20 Ibid., p. 247.
21 Ibid., p. 249.
22 Michael McCarthy, 'Bird Census Exposes 75-Year Demise of Sparrows', *Independent*, 6 November 2000.
23 Ibid.
24 Pierandrea Brichetti et al., 'Recent Declines in Urban Italian Sparrow *Passer* (*domesticus*) *italiae* Populations in Northern Italy', *Ibis*, CL (2008), pp. 177–81.
25 Ibid.
26 Rama Devi Menon, 'A Voice for the Sparrow, *The Hindu*, 13 March 2010.
27 Bryan Walsh, 'Heroes of the Environment', *Time International* (South Pacific Edition), 6 October 2008, pp. 24–5.
28 Chris Buckley, 'Beijing Journal; From Pest to Meal: A Leap Forward?' *New York Times*, 3 April 2002, p. A4.
29 At www.bto.org/appeals/house_sparrow.htm, accessed 21 April 2009.
30 Nigel Reynolds, 'Dying Sparrow Ruffles Feathers in Name of Art', at www.telegraph.co.uk/news/uknews/1461598/Dying-sparrow-ruffles-feathers-in-name-of-art.html, accessed 17 February 2011.
31 'Mystery of the Vanishing Sparrow', *Independent*, 2 November 2008.
32 Kate Vincent, 'Investigating the Causes of the Decline of the Urban House Sparrow *Passer domesticus* Population in Britain', PhD thesis, De Montfort University, Leicester (2005).
33 Denis Summers-Smith, 'House Sparrows Down Coal Mines', *British Birds*, LXXIII (1980), pp. 325–7.
34 Kate Vincent, interview with the author (January 2010).
35 Ibid.

36 Sarah Mukherjee, 'Making a Garden Sparrow-Friendly', at http://news.bbc.co.uk/2/hi/sci/tech/7739645.stm, accessed 17 February 2011.
37 Christopher Bell, interview with the author (September 2010).
38 Rob Lensink, 'Range Expansion of Raptors in Britain and the Netherlands Since the 1960s: Testing an Individual-Based Diffusion Model', *Journal of Animal Ecology* LXVI/6 (1997), p. 818.
39 Bell, interview with the author.
40 Denis Summers-Smith, quoted in Ted Anderson, *Biology of the Ubiquitous House Sparrow* (Oxford, 2006), p. 296.
41 Bell, interview with the author.
42 Ted Anderson, *Biology of the Ubiquitous House Sparrow* (Oxford, 2006), p. 299.
43 Joyce J. Persico, 'Public Outcry Saves Birds', *Times* (Trenton, NJ), 11 July 2008.
44 Roger Catlin, 'Sculptor's UConn Show Ruffles Feathers', *Hartfort Courant*, 23 April 2009, p. 12.
45 Elizabeth Bishop, quoted in Susan McCabe, *Elizabeth Bishop Her Poetics of Loss* (University Park, PA, 1994), p. 106.
46 Ibid., p. 246.

Select Bibliography

Anderson, Ted R., *Biology of the Ubiquitous House Sparrow* (New York, 2006)

Applegate, Debbie, *The Most Famous Man in America: The Biography of Henry Ward Beecher* (New York, 2006)

Baptista, Luis Felipe, *Song Dialects and Demes in Sedentary Populations of the White-Crowned Sparrow (Zonotrichia leucophrys nuttalli)* (Berkeley, CA, 1975)

Barrows, W. B., *The English Sparrow (Passer Domesticus) in North America* (Washington, DC, 1889)

Beadle, David, and James Rising, *Sparrows of the United States and Canada* (Princeton, NJ, 2003)

Bumpus, Hermon Carey, Jr, *Hermon Carey Bumpus, Yankee Naturalist* (Minneapolis, MN, 1947)

Coleman, Sydney H., *Humane Society Leaders in America* (Albany, NY, 1924)

Cutright, Paul Russell, and Michael J. Brodhead, *Elliott Coues, Naturalist and Frontier Historian* (Champaign, IL, 2001)

Fish, Stanley Eugene, *John Skelton's Poetry* (New Haven, CT, 1965)

Gentry, Thomas G., *The House Sparrow at Home and Abroad* (Philadelphia, PA, 1878)

Grimm, Jacob, and Wilhelm Grimm, *The Brothers Grimm: The Complete Fairy Tales* (Ware, Hertfordshire, 1997)

Grinnell, Joseph, *Report on the Birds Recorded during a Visit to the Islands of Santa Barbara, San Nicholas and San Clemente in the Spring of 1897* (Pasadena, CA, 1987)

Gurney, J. H., *The House Sparrow* (London, 1885)
Long, John L., *Introduced Birds of the World* (New York, 1981)
Mitford, A. B., *Tales of Old Japan* (London, 1888)
Morton, Martin, *The Mountain White-crowned Sparrow: Migration and Reproduction at High Altitude* (Camarillo, CA, 2002)
Opie, Iona, and Peter Opie, *Oxford Dictionary of Nursery Rhymes* (Oxford, 1951)
Ralston, W.R.S., *Russian Folk-Tales* (New York, 1873)
Seton, Ernest Thompson, *Lives of the Hunted* (New York, 1901)
Shapiro, Judith, *Mao's War Against Nature* (Cambridge, 2001)
Shonagon, Sei, *The Pillow Book*, trans. Arthur Waley (Whitefish, MT, 2005)
Summers-Smith, J. Denis, *The House Sparrow* (London, 1963)
—, *The Sparrows* (Calton, Staffordshire, 1988)
Todd, Kim, *Tinkering with Eden: A Natural History of Exotics in America* (New York, 2001)
Vincent, Kate, 'Investigating the Causes of the Decline of the Urban House Sparrow *Passer Domesticus* Population in Britain', PhD thesis, De Montfort University, Leicester (2005)
Walters, Mark Jerome, *A Shadow and a Song* (Post Mills, VT, 1992)

Associations and Websites

The Audubon Society is devoted to bird conservation in the United States.
www.audubon.org

The British Trust for Ornithology conducts research into bird ecology and raises money for, among other causes, the conservation of house sparrows.
www.bto.org

Celebrate Urban Birds is a project of the Cornell Laboratory of Ornithology. The house sparrow is one of the species used to engage the public in the bird life of cities.
www.birds.cornell.edu/celebration

Housesparrow.org, Kate Vincent's website, focuses on her research into the urban house sparrow's disappearance in Britain.
www.housesparrow.org

The Migratory Bird Center of the Smithsonian National Zoological Park is a good site for basic information about New World sparrows.
http://nationalzoo.si.edu/SCBI/MigratoryBirds/

The Royal Society for the Protection of Birds is a bird conservation organization in the UK investigating the vanishing house sparrow.
www.rspb.org.uk

Sialis, a site for bluebird enthusiasts, makes an argument for treating house sparrows as a non-native invasive species.
www.sialis.org

Acknowledgements

I would like to thank all the sparrow experts who shared their knowledge, particularly those who let me interview them about their work: Creagh Breuner, Andrea Crino, Teegan Docherty, Brett Klaassen van Oorschot, Denis Summers-Smith, Christopher Bell and Kate Vincent. I am also grateful to Kirstin Mattson, Felice Fischer and Yasuko Tsuchikane for information about sparrow art and the Endowment Fund of the School of Humanities and Social Sciences at Penn State Erie, The Behrend College, for assistance in procuring it. I was greatly helped in editing by Ian Blenkinsop, Michael Leaman and Jonathan Burt of Reaktion Books, and the advice and editorial eyes of Erica Olsen, Tom Noyes, Peter and Gail Todd and Jay Stevens.

Photo Acknowledgements

American Antiquarian Society: pp. 57, 85, 113; Baldwin Library of Historical Children's Literature: pp. 83, 84; Bigstock: pp. 20 (Vasiliy Vishnevskiy), 21 (Cathy Keifer), 96 (Kostiuk Eugeniy); Bodleian Library, University of Oxford: pp. 48, 49; © The Trustees of the British Museum: pp. 36, 50, 59, 61 top, 69, 71, 76, 82; The Cleveland Museum of Art: pp. 46, 60; Andrea Crino: pp. 132, 135, 137; Antoli Dubkov: p. 9; Todd R. Forsgren: p. 133; Courtesy of Stephanie Frostad: p. 45; Getty Images: p. 139; Indianapolis Museum of Art: p. 154; Collection International Institute of Social History, Amsterdam: p. 64; Institute for Wildlife Studies: p. 145; Istockphoto: pp. 6 (Martha Last), 24 (Judy Ledbetter), 25 (Diane Labombarde), 26 (Wouter van Caspel), 29 (Hansjoerg Richter), 130 (Jim Nelson), 156 (Jet Chen Tan); Lisa Jolin: p. 19; Michael Leaman: p. 63; Library of Congress, Washington, DC: pp. 22, 38, 88, 114; Courtesy of Neeta Madahar: p. 27; Collezione Sir Denis Mahon. In deposito presso la Pinacoteca Nazionale di Bologna: p. 47; Image © The Metropolitan Museum of Art: p. 10; Nature Forever Society: p. 150; Niebrugge Images: p. 163; Reproduced with the permission of Rare Books and Manuscripts, Special Collections Library, the Pennsylvania State University Libraries: p. 98; Philadelphia Museum of Art: p. 61 bottom; Rex Features: pp. 66 (Roger-Viollet), 78 (Stuart Atkins), 149 (Alisdair Macdonald), 157 (Roger-Viollet), 158 (Canadian Press); Shutterstock: p. 23 (Wrangler), 124 (Marin Trajkovksi), 147 (Igor Golovniov); Jason Sullivan: p. 12; Kim Todd: p. 160; University of California, San Diego: p. 70; U.S. Fish and Wildlife Service: pp. 18 (Thomas G. Barnes), 30, 33 (Dave Menke), 34 (Lee Karney), 41, 129 (Gary Kramer),

140 (P.W. Sykes); U.S. Navy: p. 13 (Mass Communication Specialist 2nd Class Kristopher Wilson); Victoria & Albert Museum, London: pp. 37, 94.

Index